AMBASSADORS IN BLUE

IN EVERY CLIME AND PLACE

The Marine Security Guard Program

MSgt Andrew A. Bufalo USMC (Ret)

ISBN 978-0-9845957-3-0

First Printing – February 2011
Printed in the United States of America

www.AllAmericanBooks.com

Ambassadors In Blue

OTHER BOOKS BY ANDY BUFALO

SWIFT, SILENT & SURROUNDED
Sea Stories and Politically Incorrect Common Sense

THE OLDER WE GET, THE BETTER WE WERE
MORE Sea Stories and Politically Incorrect Common Sense

NOT AS LEAN, NOT AS MEAN, STILL A MARINE
Even MORE Sea Stories and Politically Incorrect Common Sense

EVERY DAY IS A HOLIDAY, EVERY MEAL IS A FEAST
A Fourth Book of Sea Stories and Politically Incorrect Common Sense

THE ONLY EASY DAY WAS YESTERDAY
Fighting the War on Terrorism

TO ERR IS HUMAN, TO FORGIVE DIVINE
However, Neither is Marine Corps Policy
A Book of Marine Corps Humor

HARD CORPS
Legends of the Marine Corps

SALTY LANGUAGE
An Unabridged Dictionary of Marine Corps Slang, Terms & Jargon

THE LORE OF THE CORPS
Quotations By, About & For Marines

"Eternal vigilance is the price of liberty."

– Wendell Phillips, American Orator; 1811-1884

Dedicated to the Marine Security Guards who have died in the line of duty while protecting our diplomatic missions around the world.

Corporal James C. Marshall, Saigon, 1968
Sergeant Charles W. Tuberville, Pnohm Penh, 1968
Corporal Darwin D. Judge, Saigon, 1975
Corporal Charles McMahon Jr., Saigon, 1975
Sergeant Bobby A. Romero, Paris, 1978
Corporal Stephen J. Crowley, Islamabad, 1979
Corporal Robert V. McMaugh, Beirut, 1983
Corporal Thomas Smith, Beirut, 1983
Staff Sergeant Bobby J. Dickson, San Salvador, 1985
Staff Sergeant Thomas R. Handwork, San Salvador, 1985
Sergeant Patrick R. Kwiatkowski, San Salvador, 1985
Sergeant Gregory H. Weber, San Salvador, 1985
Sergeant Jesse N. Alingala, Nairobi, 1998

Ambassadors In Blue

PREFACE

Duty as a Marine Security Guard is one of the most unique and prestigious things a Marine can do during his time in the Corps, and anyone who stands post at an American diplomatic mission overseas is in essence manning the front lines of freedom during both times of war and peace. Not only do they do a superlative job of ensuring the security of our embassies and consulates, they also play a significant role in the conduct of American diplomacy.

A young Marine, for instance, was the first American Stalin's daughter confronted when she defected. She simply walked into the American Embassy in New Delhi, India after office hours. On duty was a Marine Security Guard, and it was to him that she made her first appeal for help from the United States. Marines standing post have careful instructions to call in experienced embassy officers in case of such emergencies, but even so a tremendous responsibility fell upon the shoulders of that young Marine.

Imagine for a moment the dilemma he suddenly confronted. A woman had appeared without warning out of the night, claiming to be Stalin's daughter - not a very probable story, since embassies attract plenty of eccentrics. If this were a prankster or crank he would have to jolly her along and try to get rid of her. On the other hand if this were *really* Stalin's daughter, her defection would be tremendously important to the United States and he could not risk antagonizing her. He had to display just the right blend of sympathy and reserve, because he did not want her to panic and flee before embassy officers could arrive. At the same time, neither could he be so forthcoming that she could later allege he had made commitments to her on behalf of the United States. That young Marine was the American who had to make the first - and some of the key - decisions which led to the startling propaganda coup of Stalin's daughter's defection to the United States.

From out of the night another enormous challenge confronted other Marine Security Guards in Saigon. At 0250 on January 31, 1968 the Viet Cong attacked the American Embassy there. Two Marines were on duty in the lobby when they heard the first shots, and they pulled the unarmed Vietnamese front gate guard from certain death and slammed shut the three-inch teakwood front doors. With bullets ricocheting through the lobby they coolly telephoned for reinforcements and broke out additional weapons, and as they did a rocket came through the wall and exploded. One Marine was seriously wounded and the other slightly. Rockets continued to explode on the embassy's facade. As reaction forces arrived the Viet Cong attack on the building itself diminished, but their fire was so intense that the American reinforcements could not get into the embassy itself. The two Marines, along with another Marine stationed on the roof and the few embassy staff members on duty, were without physical contact with other Americans through most of the night. A helicopter got through to the embassy roof at about 0615 to evacuate the seriously wounded Marine, but the intense fire caused the helicopter to make an emergency landing in a rice paddy before the Marine could be taken to a hospital. Only after first light, about an hour later, could 101st Airborne Division paratroopers land on the roof and fully secure the building. Until then, the courage and coolness of two Marines had played a vital role in denying the Viet Cong entrance into the embassy itself.

During the night of 25 January, 1968, in another part of

the world, the Marines on duty at the embassy in Panama learned the building's roof was on fire. The Marines found the top door engulfed in smoke and secured the char force - but one man was missing. He was trapped in the elevator. Using a fire ax, the Marines forced open the elevator door and got the man out safely. Meanwhile the fire department, along with off-duty watchstanders and appropriate embassy officers, had been notified. The firemen put out the fire while American personnel maintained control over the areas containing classified material. The Marines' calm efficiency and rapid performance had led to early containment of the fire and prevented large-scale damage to the embassy.

Fire also created a sudden emergency in Kinshasa, in what is now known as The Democratic Republic of the Congo. On 11 April, 1968 a woman ran into the Embassy to tell the Marine on duty that the Spanish Embassy next door was on fire. The Marine immediately summoned the fire department, the NCOIC, and off-duty watchstanders. Leaving one Marine to maintain security at the embassy, the others rushed to extinguish the blaze. The door was secured, but one Marine immediately scaled the wall and found a second story window open. Another Marine followed right behind and tried to pass him a fire extinguisher, but in doing so he grabbed an electrical wire for balance. The wire was live, and the jolt threw the Marine to the ground - fracturing a bone in his left foot. Another Marine succeeded in passing the extinguisher to the first Marine, who then managed to put out the fire. In the end, as a result of the Marines' quick thinking and decisive action, both the Spanish and American Embassies were spared extensive damage.

Regularly, in many less crucial or heroic ways, Marine Security Guards play a personal part in the day-to-day details of American Embassy operation. It is the Marine on duty who receives the often-anguished request of an American citizen seeking emergency help at night or on a weekend. He has to be a model of tact and assurance until he can sort the problem out and contact the embassy's duty officer. Many less urgent requests come to him for street directions, locations of hotels or restaurants, sightseeing, etc. and he serves as a "short-order" information service, dispensing answers in a reassuringly American accent. In similar fashion, he deals with foreign citizens who come to the embassy seeking information. The stream of such callers is heavy on American holidays or on Saturdays, which are normal workdays in many countries. To the skills of tact and competence, he must often add some knowledge of the local language.

The help extended by Marine Security Guards often goes beyond the strict limits of duty. Many an American who has suddenly found himself without funds overseas has been comforted by a cup of coffee, a sandwich or some cigarettes provided by an MSG. On occasion a Marine has even kept a benevolent eye on a penniless American sitting out the night in an embassy lobby until help can be arranged in the morning, and countless times Marines have helped frightened, bewildered or belligerent Americans with soothing conversation and patient comfort. A trim and efficient Marine at the reception desk is a familiar and reassuring sight for visitors to American diplomatic missions overseas.

The Marine Security Guard program has been invaluable to the U.S. Foreign Service. It has provided American diplomatic and consular missions with highly competent protection for their classified material and assistance in protecting government property and the lives of employees. Prior to the program's inauguration the Department of State had had to hire civilians, American and foreign, to protect its establishments abroad. Only a limited number of guards could be hired, and often they were of doubtful background, ability and suitability – and in many cases the guard positions were found to attract only the old and the lazy. Furthermore, many of the American guards resided permanently in foreign countries and were married to aliens. This system was obviously unsatisfactory in the rising international tensions which followed WWII, and thought naturally turned toward a guard force which was young, alert, well trained and had military discipline.

The historic association between the U.S. Foreign Service and Marine Corps suggested the Marine Security Guard program. Throughout the history of the United States Marines had served many times on special missions as couriers, guards for embassies and legations, and to protect American citizens in unsettled areas such as China and Cuba. Probably the most dramatic example was the defense of the legation in Peking in 1900 against the siege by the Chinese "Boxers." In fact President Roosevelt, at the request of the Department of State, issued a Presidential Order establishing guards at certain embassies and legations. Drawing on this precedent, and with great foresight, the drafters of the 1946 Foreign Service Act included authorization for the Secretary of the Navy, on request from the Secretary of State, "to assign enlisted men of the Navy and the Marine Corps to serve as custodians" at embassies, legations and consulates.

The Marine Security Guard program is now more than sixty years old. A Memorandum Agreement was signed on December 15, 1948 by Secretary of the Navy Sullivan and Undersecretary of State Lovett, and the first eighty-three Marines reported to the Foreign Service Institute of the Department of State in January of 1949 for training. Then, on January 28, 1949, the first fifteen Marines departed for assignment to Foreign Service posts abroad - six went to Bangkok, and nine to Tangier. The program developed rapidly, and by the end of May 1949 over three hundred Marines had been assigned to posts around the world. The agreement covering the Marine Security Guard program was later renewed and broadened in 1955 with the signature of a "memorandum of agreement between the Department of Defense and Department of State."

To facilitate administration of its personnel, Headquarters United States Marine Corps established Company F to perform administrative and training functions for the program, and in 1952 a junior Marine officer was assigned to each of the regional security headquarters of the Department of State at Frankfort, Beirut, Manila and Panama. Fox Company was upgraded in February of 1967 to become the Marine Security Guard Battalion, and at that time Headquarters Company was established at Henderson Hall while Company A served the embassies in Europe, Company B guarded the American missions in Africa and the Middle East, C Company took care of the Far East, and embassies in South America belonged to Company D.

The composition of Marine Security Guard Detachments and their duty situations vary. At any minute of any day Marines are standing duty at embassies around the world, and while most detachments consist of about six-to-eight Marines, the larger or more sensitive posts have more. Each detachment is headed by a staff non-commissioned officer now known as the Detachment Commander, but formerly the Non-Commissioned Officer in Charge or NCOIC. The DetCmdr is on duty at the embassy during normal working hours and at such other times as necessary to supervise the work of the detachment, and at most posts one additional

Marine is on duty twenty-four hours a day manning "Post One." While his primary function is to control access to the embassy he is also ready to help callers, receive emergency messages and deal with any other problems which may arise. After hours this Marine tours the building periodically to be sure all classified material has been properly secured and escorts the char force (janitors) while they clean classified areas, in addition to handling other duties.

Marine Security Guards wear uniforms while on duty at most posts and civilian clothing the rest of the time, and before departing for their first assignment are issued a set of dress blues and outfitted with appropriate civilian clothing at Department of State expense. Quarters are also provided by the Department of State, and usually consist of a house or apartment for the Detachment Commander and a large communal house for the other members of the detachment.

Requirements for the program are simple but rigorous. A Marine must be single if a sergeant or below and remain single until his tour is completed. He must be a lance corporal or above, have had at least eighteen months active service and thirty months or more left to serve. Corporals and lance corporals must have average proficiency and conduct marks of 4.4 or higher, and all applicants must U.S. citizens, in excellent physical condition, and be able to qualify for a top secret security clearance.

Candidates undergo an intensive program of instruction and two thorough screenings, during which a quarter or more are eliminated. The principal subjects covered include physical security, protocol, the Foreign Service, security tasks, dealing with hostile demonstrations and mob action and countering measures of espionage and subversion. Once the classroom work is done the students are taken to the State Department itself where they try to find security violations in the building. Once overseas, Marines must also study foreign languages for at least one hundred hours where applicable. Most acquire some knowledge of the local language, and many become quite proficient.

The Marine Security Guard program has assumed added importance because of the increased frequency of terrorist attacks, mob action or other attacks against American Foreign Service posts. During sensitive periods MSGs inspect all packages entering an embassy and carefully check restrooms and other public areas for planted explosives, and dealing with mob actions requires not only courage but careful judgment and firm discipline. The integrity of the building and files must be protected if at all possible, but actions must be avoided which would endanger American lives - particularly those of defenseless American wives and children in their homes. MSGs can also fulfill a damage-control function. Prompt and courageous use of fire hoses by Marines prevented extensive damage during an attack on the American Embassy in Cairo in 1961, and they did so in a hail of stones from the mob which resulted in one Marine being badly gashed over one eye. The presence of armed and obviously competent Marines may have also deterred the out-and-out sacking or further degradation of the Foreign Service posts which were attacked, and certainly the presence of Marines inside embassy compounds has often been psychologically comforting to American personnel during other hostile demonstrations and civil disturbances.

At the same time Marine Security Guards have compiled an amazing record in fostering better relations with people all over the world. It is a trite but true comment that each Marine abroad is personally an ambassador of the United States, because by his behavior foreigners judge Americans. He is viewed especially as an example of both American youth and military forces. Marine detachments have done a wonderful job by simply reflecting the warmth, generosity and initiative of the American people, and in their free time they have carried out an untold number of projects to help the people amongst whom they live.

A sample of those projects demonstrates vividly the concern and initiative of the Marines. The detachment at the embassy in New Delhi, India sponsored a charity ball for a hospital, earlier manned a fruit stand to raise money for a sterilizer, and also participated in a fund raising drive for retired Indian soldiers. The detachment in Vientiane, Laos once constructed swings and seesaws for the children of the Catholic orphanage, in Bangkok the MSGs held a series of dinner parties for the Marines of Thailand, and the detachment in Abidjan, Ivory Coast conducted classes in physical training and body building at the Ivory Coast University. The Manila detachment reconditioned a bus no longer needed by the U.S. Aid program and presented it to an orphanage, and not only was the bus put into first class running condition, it also sported new tires and a newly painted body. In Colombo, Ceylon the entire detachment responded to a call by the hospital ship Hope for blood donations due to insufficient local supplies, the Marines in Dhahran, Saudi Arabia rounded up magazines for the local eye clinic and hospital, and aid and cheer were brought to a Madrid school for homeless and destitute children when the Marine detachment there held a Three Kings Day party for the girls and donated badly needed supplies and cash to the school. This list of projects - covering only a few months, chosen at random, and far from complete - speaks for itself.

Understandably, the Foreign Service and many other Americans abroad have a warm spot in their hearts for embassy Marines. Marine House, the quarters for the detachment, is invariably a well-known locale which everyone at the embassy from the ambassador on down visits, and the annual Marine Birthday Ball is invariably the highlight of the social calendar at every post lucky enough to have a Marine Detachment.

The Marines also participate in embassy and local American community life, playing in softball leagues in Geneva, Kabul, New Delhi, Tel Aviv, Calcutta, Dhahran and Helsinki. They also have played Volleyball in Nicosia and Bombay and basketball in Ankara, organized a bowling league in Tel Aviv, and even sponsored a team in a dart league in New Delhi. Marines have coached Little League teams in Rabat and Ankara and a football team at the American School in Tehran. In Amman they conducted physical education classes at the American Community School, and in Ankara supervised an annual summer physical fitness program. Marines have led Scouting activities in Amman, Bogota, Mexico City and Tegucigalpa, and during the long winter in Reykjavik showed full length movies Sunday afternoons and evenings. In Ankara, Turkey the annual Easter egg hunt was even held in the Marine House yard.

Marines also participate actively in American community endeavors. Around the world they routinely play an honored and helpful role in celebrations of the Fourth of July, and in Wellington, New Zealand members of the detachment took part in an annual ceremony at the memorial honoring Admiral Richard Byrd, explorer of the Antarctic. In Tripoli, Libya the detachment has participated in the Memorial Day ceremony at the graves of five American seamen killed in the explosion of *USS Intrepid* during the Barbary Wars, and the Marine detachment in Tokyo once found the strength to fulfill the request of the press attaché and escort Miss

Universe to a reception in her honor.

The preceding should give you a basic idea of the perks - and responsibilities - associated with being a Marine Security Guard. The purpose of this book is to give Marines considering such duty an idea of what to expect, those on the program some tips for a successful tour, and everyone else some historical context and a glimpse of what goes on behind the scenes. Naturally things change from time to time, such as the re-designation of the Marine Security Guard Battalion as the Marine Embassy Security Group to reflect its place in a changing world (although the old MSG Battalion shield was used on the cover to pay homage to the "old guard"), but the basic mission and overall flavor of the duty has changed little since the very first MSG took his post at an American Embassy overseas and earned the title "Ambassador in Blue."

Ambassadors In Blue

TABLE OF CONTENTS

HISTORY, BACKGROUND & ORGANIZATION

HISTORY OF THE MSG PROGRAM

Shortly after World War II, the Department of State decided to re-examine the problem of obtaining sufficient guards of appropriate caliber for the protection of Foreign Service posts. Prior to this time, the Department had followed the practice of hiring civilians, both American and foreign, for the protection of its missions abroad - a practice which in many cases had been proven unsatisfactory. The problem was based in the inadequacy of the civilian guard salary scale, which permitted only a limited number of guards of doubtful background, ability, and suitability. In many cases the guard positions were found to attract only the old and the lazy, and many of the American guards resided permanently in foreign countries and were married to aliens. Thus, in a period of growing international tension such as existed in 1947, it was only natural that thoughts should turn toward the establishment of a guard force which was young, trained, and under strong discipline – in other words, a military force.

Early in 1947, an informal approach was made to the War Department to see if military guards might be supplied to Foreign Service posts. Although some interest was shown toward the proposal, substantial cuts in military appropriations were being implemented at the same time, so no plan for military guards seemed feasible. However, on September 8, 1947, the proposal was again made that the War Department furnish military personnel for Foreign Service guard duty, with additional provisions which would relieve the War Department of any unusual expense. The general plan advanced was that the War Department would pay only the cost of basic salaries, uniforms and equipment, and the State Department would pay the cost of transportation, rentals and cost of living. While it was estimated that the State Department could obtain three times as many guards as could be supported by appropriations for civilian guards, it was also pointed out that the War Department could use the attraction of world-wide service as a recruiting inducement. It was further suggested that a working committee of State Department officers be appointed to discuss the matter with the appropriate War Department officials.

However, before any further steps were taken in this direction, it was correctly pointed out by the Legal Advisor to the Secretary of State that, under section 562 of the Foreign Service Act of 1946, the State Department could not enter into any agreement concerning guards with the War Department, but such action *could* be taken jointly with the Department of the Navy. As the Act stated:

"Sec. 562. The Secretary of the Navy is authorized upon request of the Secretary of State, to assign enlisted men of the Navy and Marine Corps to serve as custodians under

supervision of the principal officer at an embassy, legation or consulate."

With this interpretation, it was clear that the Department of State would thereafter have to cooperate with the Department of the Navy and the Marine Corps in order to bring an efficient security guard force into the Foreign Service. However, it should be pointed out that the association between the Marines and the Foreign Service was not altogether new. Throughout the history of our country Marines had served many times on special missions as couriers, guards for Embassies and Legations, and to protect American citizens in troubled areas such as China and Cuba. In fact President Theodore Roosevelt, at the request of the Department of State, issued a Presidential Order just after the turn of the century which established guards at certain Embassies and Legations similar to those we have today. Thus, the provisions of Section 562 of the Foreign Service Act of 1946 actually served the purpose of renewing and re-invigorating an association which had long been performed ably and well in the service of American interests abroad.

Next began the series of discussions between the two Departments which led to the formal establishment of the Security Guard Program. In February and March of 1948, budgetary and personnel limitations on the Foreign Service Staff resulted in an increased pressure toward the completion of an agreement on Marine Guards, and surveys of the posts in the field were made to ascertain the number of Marines that would be needed. At this time it was contemplated that Marines might be assigned only to the less troubled areas of the world, thereby freeing civilian guards for duty in more sensitive posts. In April of 1948 the Marine Corps agreed to participate in the program, with the general provisions that no Marines would be assigned to areas primarily under Army jurisdiction, that Marines would be assigned directly to the Naval Attaché at each post, and that a fair difference be made between the Marines' salary and the civilian guards' pay. After opinions had been solicited from the various areas of the State Department, these general principals were submitted to the Under Secretary of State, Mr. Robert A. Lovett, who agreed to them on May 26, 1948. Mr. Lovett presented a request to the Secretary of the Navy, Mr. John L. Sullivan, on June 22, 1948, asking that initially three hundred Marines be assigned to duty abroad on Foreign Service guard detachments. These first Marine Security Guards were to be assigned as members of the staffs of Naval Attachés, with the Navy being responsible for administration and discipline and the State Department agreeing to pay for salary compensation, allowances, travel, and for civilian clothing when necessary. Representatives from both Departments were designated to work out the details, and by July 20, 1948 assignment of the initial three hundred Marines was authorized by Acting Secretary of the Navy John N. Brown. These Marines were the assigned to the field based upon the security needs of the posts.

It was still necessary to negotiate a formal agreement between the two Departments pertaining to the use of the Marines at Foreign Service posts. Preliminary discussions were held during the months of August and September which resulted in a Tentative Agreement reached on September 20, 1948. Some of the high points of this agreement were: Marines were to be responsible to the Chief of Mission or Principal Officer of the Foreign Service Post, through the Senior Marine Commissioned or Non-Commissioned Officer and the Naval Attaché, where assigned; salaries were to be paid by the Marine Corps, but the State Department assumed obligation for allowances when government facilities were not available; also, the Marines were to utilize government transportation when it was available, but the State Department would arrange for transportation when it was not. On the basis of these first three hundred Marines, it was estimated that the United States would realize a savings of $160,750.00 per year over the employment of a similar number of civilian guards, with the added advantage of having young, trained guards under military discipline. October 25, 1948, was set as the target date for having Marines in the field, and it was planned that they would be provided with five to ten days of training before being sent out.

An incident which arose from these early discussions concerned the matter of whether Marine should in all cases serve in uniform or could, when politically expedient, wear civilian clothing. It was agreed that the Marines would serve in uniform whenever possible, but the State Department asserted that certain circumstances might exist (the example given was turbulent Cairo) which would make wearing civilian clothing mandatory. This matter came to the attention of the President himself, and was resolved on November 5, 1948 when President Truman approved a plan whereby Marine detachments in civilian clothes could be assigned to posts when the Secretary of State determined that circumstances warranted. This policy on the wearing of uniforms is still in effect, although the tendency in recent years has been more and more toward enabling the Marines to wear uniforms while on duty in Foreign Service posts.

Through the fall of 1948, other details were ironed out through further discussion between the two Departments. The State Department assumed full responsibility for medical care of Marines at the posts and established a civilian clothing allowance of $300.00 for temperate zone posts and $239.00 for tropical zone posts. Specific assignment of individuals became the joint responsibility of the State Department and the Marine Corps, with the State Department developing the overall placement schedule. Preliminary to drawing up implementation plans, the chiefs of mission at posts around the world were queried as to whether Marines would be welcome and useful as guards at their posts. When the indications were that Marines would be an asset to post security, permission to use them as security guards then had to be sought from each host government concerned. On November 29, 1948, after this and other groundwork had been laid, the Department of State finally presented to the Secretary of the Navy a draft Memorandum of Agreement between the two Departments on the subject of Marine Guards. Secretary of the Navy Sullivan signed the Agreement on December 15, 1948, Under Secretary Robert A. Lovett signed for the Department of State, and this most basic document, known as the "1948 Memorandum of Agreement," came into force.

With the signing of the formal agreement, the next task was to set up the necessary machinery for administration of the program. Meanwhile, it was important for the Marines to reach their posts of assignment as soon as possible, and this project was attacked with enthusiasm. In early January, 1949 eighty-three Marines were assigned to the Foreign Service Institute for training, and on January 28, 1949, the first fifteen Marines departed Washington on their assignments, six for Bangkok and nine for Tangier. A letter of Instruction to the field, coordinated between the Marine Corps and the Regional Bureaus of the Department of State, was then circulated in order to outline the uses to be made of the

8

Marines in the Foreign Service. Also, to facilitate the administration of its personnel, Headquarters, United States Marine Corps expanded Casual Company (later re-designated Company "F") of its Headquarters Battalion in order to provide for its personnel on separate duty, of which Marine Security Guards constituted the majority. The program developed rapidly, and by the end of May, 1949 303 Marines were assigned as Foreign Service guards at posts throughout the world.

As might be expected, it was necessary to contend with many problems of an administrative nature during the first years of the program. One question concerned whether Marines who married while at their posts abroad, whether to aliens or to American citizens, should be retained at their posts. On the basis of excessive costs for security screening, housing, and transportation of new dependants, and in order to ensure an effective mobility of the Guard Corps when its services were urgently required, a general policy of replacing Marines who married while abroad with unmarried Marines was adopted. Similarly, it was determined that Marines who became involved in serious breaches of discipline, or who engaged in actions detrimental to the interests of the United States, would be returned home for disciplinary or other action. These policies were promulgated by the Commandant of the Marine Corps in September, 1949. It was also necessary to clarify the position of the Marine Guards within the ordinary chain-of-command at Foreign Service posts. To this end, a joint declaration was issued which stated that it was the intent of the State Department and Marine Corps that Marine Guards would serve directly under the Security Officer of the post for all duty and administration. The role of the Naval Attaché was to be that of a senior officer of the Naval Service present, with no power to exercise control over Marine personnel at the post beyond admonition or, in serious cases, forwarding of complaints to the Marine Corps for action. These and many similar problems were resolved during the first year, and the program was performing its duties with an efficiency heretofore unknown in the Foreign Service.

But growing pains were felt before the Security Guard Program had celebrated its first anniversary. With the inception of the Military Defense Assistance Program late in 1949, it was estimated that forty additional Marines would be required to carry out guard duties at eight MDA missions in Europe. In addition, it was felt the Foreign Service was critically in need of an increase in Marines as replacements for civilian guards in several area where the needs of the mission had changed. On January 21, 1949 the Secretary of the Navy was asked to provide forty Marines for the purpose of implementing the Mutual Defense Assistance Program, and it was also indicated that a further request might be made to increase the overall number of Marines serving at Foreign Service posts. Unfortunately, the Marine Corps' personnel ceiling limitation, coupled with a provision of the Mutual Defense Act which prohibited payment of military salaries from funds allocated to that program, presented a major obstacle to the increase in the Guard Corps at that time. However in March of 1950 the importance of the Mutual Assistance Program was felt to be so great that the State Department decided to proceed with the assignment of the forty MDAP Guards on the basis that the State Department would completely reimburse the Marine Corps for the cost of basic salaries for these and any subsequently assigned Marines. With this understanding, the forty MDAP Guards were ordered to their European posts, and by June of 1950 they were joined by twenty-four more Marines whom the

Secretary of State had requested, on a completely reimbursable basis, for ordinary Foreign Service guard duty. Shortly thereafter, with the outbreak of hostilities in Korea, the Marine Corps entered a period of rapid expansion which so enlarged the scope of Marine Corps operations that by August of 1950 it was determined that the cost of basic salaries for Marines in excess of the initial quota of three hundred could be borne by the Marine Corps alone.

During the fall of 1950 the Department of State, the Marine Corps and the Bureau of the Budget participated in a series of discussions to determine what should be the exact relationship between the Department of State and the Marine Corps on the question of financial support of the Marine Guards. At the conclusion of these discussions the Bureau of the Budget ruled: "If the guarding of American missions abroad is essential to the security of the United States, then that guarding is a defense function and should be carried out by the Military Establishment without reimbursement of basic costs." The Department of Defense was then asked to accept responsibility for the guarding of Foreign Service posts abroad and to assign that duty to the Marine Corps. It was recommended at the same time that the Marine Corps be allowed to increase its personnel ceiling by 650 enlisted positions in order to provide for the Marine Guards thought to be necessary for completing an effective security defense. These additional responsibilities were quickly assumed by the Department of Defense and Marine Corps, and on February 15, 1951 the Commandant of the Marine Corps ordered his Director of Personnel to increase the number of Marines available for duty with the Foreign Service by 311, to a total of 675 guard positions. Thus, at a time when the Korean War was at its height, the Marine Corps obligingly allocate 311 of its best-qualified men for duty with the

Foreign Service - thereby demonstrating that the Marines respected the request as being of a vital and urgent nature.

During the rest of 1951, the Marine Security Guard Program underwent further improvement. It was specified at this time that volunteers were preferred. Qualifications were established which required two years of satisfactory prior service with no convictions by general or special court-martial and not more than one summary court-martial within the past two years, being unmarried and agreeing to remain so, and having mature judgment, stable character and an appropriate security clearance. It was also decided to permit the assignment of married Non-Commissioned Officers-in-Charge, in order to improve discipline by obviating overly close personal contact between Guards and their NCOIC's during off-duty hours. The office of Security of the Department of State, since it had been charged with the responsibility of directing the security program, was designated as the contact point within the Department for all Marine Security Guard matters.

In 1952 an important step was taken toward improving supervision of the Guards in the field. The Marine Corps suggested a plan in which a junior Marine officer would be assigned to each Regional Security Headquarters of the Department of State to work with the Regional Security Supervisors on all matters pertaining to the Marines. This plan was soon approved, and in June of 1952 Marine officers were authorized to serve in Paris, Cairo, and Manila. These officers, to be known as Regional Marine Officers, were directed to supervise the morale, discipline, administration, and other affairs of the Marine Guards, and to make semi-annual inspections of the guard detachments within their areas of responsibility. Since the Foreign Service Act of 1946 specified that only enlisted men could be used in such programs, the Marine Corps undertook to pay all costs of the Regional Officers, a step taken readily in the interest of improving the quality of the guard force. With the assignment of an officer to Rio de Janeiro in November of 1952, the Regional Marine Officer system provided world-wide coverage.

During the last months of 1952, a poll of various posts abroad and Bureaus in Washington indicated that at least 125 more Marines were required. Aside from purely Foreign Service requests, other agencies and offices attached to the Foreign Service missions bolstered the claim for additions to the guard force. In March, 1953 the military services asked for fifty-seven additional Marines to broaden coverage for Service Attachés whose offices were located in buildings separate from the chancery buildings proper. These further requests for Marines were honored so that, by December of 1953, six officers and 676 enlisted Marines were in active support of the Foreign Service establishment, and by Fiscal year 1956, when the guards approved for extension of Attaché office protection were finally provided, the number rose to six officers and 733 enlisted men. A peak in the size of the Marine Guards was reached on August 30, 1956 when the authorized total was raised to 850 enlisted Marines, but due to reduced Marine Corps personnel authorization this number subsequently was reduced to 742 enlisted in August, 1957 - at which level it presently stands.

The problems of administering and training such an expanded body of personnel obviously became more complex. Early in 1954, the suggestion was offered that it might be to the mutual advantage of the Department of State and the Marine Corps to inaugurate an enlarged and improved training program. This program would last a minimum of thirty days and would provide Marine Corps

13

personnel assigned to Foreign Service duty with a complete general knowledge of their new assignment prior to their departure from the United States. An agreement was quickly reached for establishing a four-week Marine Security Guard Course which would train forty to fifty Marines each month, beginning in the autumn of 1954. Henderson Hall at Headquarters Marine Corps in Arlington, Virginia was designated as the training site, and the necessary instructional staff was assembled so that the first class began its training on November 4, 1954.

Requirements for Marines seeking Foreign Service duty were made more strict by a letter from the Commandant of the Marine Corps dated July 26, 1954. The new training course was set up to comply with these requirements, and except for minor changes is still operating in the following manner. Candidates for this duty are sent to Marine Security Guard School, where they are carefully interviewed by experienced Marine officers and by State Department Security Officers with a Foreign Service background to determine the candidate's suitability for this type of duty. The candidate is then given five weeks of training which covers all phases of security work and includes a general indoctrination on the Foreign Service as well as instruction on such subjects as personal conduct, Communist methods and techniques, American history, and representing the United States abroad. About half of the instruction is provided by the State Department, and half by the Marine Corps. The candidates must then pass comprehensive written examinations on all subjects taught in the course, and upon successful completion must again appear before a board of State Department and Marine officers for final screening. By this time approximately 25-40% of the candidates have been selected out of the program and returned to other Marine duties, while those who have passed all phases of the course are assigned to posts abroad.

After 1952 a great need was felt for the general revision of the "1948 Memorandum of Agreement" between the Navy and State Departments which, in the light of such great expansion of the Marine Guard, both in organization and purpose, was seen to have become too restrictive to provide effectively for the administration of the program. Representatives of the Department of State and Marine Corps had drawn up a new Draft Agreement in 1953 which was not acceptable to the Department of Defense, and the draft was retained at that level until April of 1955 when it was returned, revised but unsigned. After re-study by interested areas of the Departments of State and Navy, as well as by the Marine Corps, acceptable provisions were found which enabled the Draft Agreement to be sent again to the Department of Defense in September, 1955. Signatures by Charles E. Wilson, Secretary of Defense, and John Foster Dulles, Secretary of State, on September 22, 1955 finally made official the "1955 Memorandum of Agreement between the Department of Defense and Department of State," a badly needed enabling act which is still in effect. This modernization process was carried a step further in 1956 when the 84th Congress repealed Section 562 of the Foreign Service Act of 1946 and adopted Public Law 1028 (70-A Statutes-at-Large 374). This law restated the provisions for use of military personnel for custodial duty at Foreign Service establishments and is now to be found in codification of Title 10 USC 5983, which is the present legal basis for the Marine Security Guard Program.

Recent years have seen a continual effort on the part of the Marine Guard Program to adjust to a constantly-changing world situation with its resulting shifting of emphasis of

15

security needs. The increasing frequency of mob action at Foreign Service installations, inspired by Communists and ultra-nationalists, forced the Department of State to become more concerned with the problem of protecting American lives. Every effort was made to strengthen the physical security of posts where such actions were a possibility, but to enhance these efforts a revision was made to part of the training at the Marine Security Guard School to include briefings on the responsibility of Marines to take effective action to defend lives. To aid in general training, a "Handbook for Marine Security Guards" was published in July 1956 and distributed to appropriate personnel throughout the world. The Marine Guard also took on other tasks, such as the provision of guard service for international conferences. When notice was taken of their excellent performance at Foreign Service posts, the Marines were asked to provide the guard for the American delegation to the 10th Anniversary Conference of the United Nations at San Francisco in June of 1955. Subsequently the Marines have provided security guards for American delegations at practically every international conference that has been held. With the addition of operations such as these, the end of the first ten years of the Guard Program found the Marines involved on a world-wide basis in a variety of duties in direct support of the security needs of the United States.

The use of Marines as security guards at our Foreign Service establishments abroad has come to be accepted as normal practice, and has evoked much interest and favorable comment from the American public. The American tourist who visits many of the larger posts abroad now expects to find a capable young Marine on duty. Members of Congress and other high government officials, as well as the Department of State and the Department of the Navy, have all taken a keen interest in this program. The Commandant of the Marine Corps on one occasion stated that he held himself personally responsible to the Secretary of State for the performance of Marine Guard personnel and that he would accept nothing less than duties which were "well done." This attitude, which certainly characterized the performance of the Marines during the first ten years, assure that Marine Security Guards should continue to receive well-earned praise and respect for their important service in the interests of the security of the United States.

MARINE CORPS EMBASSY SECURITY GROUP
(Formerly MSG Battalion)

Mission

Marine Security Guards provide security services to selected Department of State Foreign Service posts to prevent the compromise of classified material and equipment and to provide protection for United States citizens and United States Government property. The Marine Security Guard Battalion exercises command, less operational control, of these Marines, in that it is responsible for their training, assignments, administration, logistical support, and discipline.

MSGs focus on the interior security of a diplomatic post's building(s). In only the most extreme emergency situations are they authorized duties exterior to the building(s) or to provide special protection to the senior diplomatic officer off of the diplomatic compound. MSGs carry a certain level of diplomatic immunity in the performance of their official duties.

The MSG Program in its current form has been in place since December of 1948, but the Marine Corps has a long history of cooperation with the Department of State going back to the early days of the Nation. From the raising of the American flag at Derna, Tripoli to the secret mission of Archibald Gillespie in California to the 55 Days at Peking, Marines have served many times on special missions as couriers, guards, and to protect American citizens abroad.

Organization

There is a joint working relationship between MSG Battalion and DOS. The Detachment Commander reports to the Battalion Commander via the Company Commander in the administrative chain of command. In the operational chain of command, he reports to the Chief of Mission via the Regional Security Officer or Post Security Officer. The joint Memorandum of Agreement governs these relationships.

The Battalion Commander reports to the Director of Operations (PO) at Headquarters Marine Corps (HQMC). MSG Battalion currently fields over one thousand Marines at 121 Detachments organized into nine regional MSG companies and located in over 105 countries.

- **Headquarters Company** and Battalion Headquarters is located aboard Marine Corps Base Quantico, Virginia. The MSG School is part of Headquarters Company.

- **Region 1** (Company A) Headquarters is located in Frankfurt, Germany and is responsible for Eastern Europe.

- **Region 2** (Company B) Headquarters is located in Nicosia, Cyprus and is responsible for northern Africa and the Middle East.

- **Region 3** (Company C) Headquarters is located in Bangkok, Thailand and is responsible for the Far East, Asia and Australia.

- **Region 4** (Company D) Headquarters is located in Ft. Lauderdale, Florida and is responsible for Central and South America and the Caribbean.

- **Region 5** (Company E) Headquarters is located in Frankfurt, Germany and is responsible for Western Europe.

- **Region 6** (Company F) Headquarters is located in Pretoria, South Africa and is responsible for Sub-Saharan Africa. This headquarters was previously located in Nairobi, Kenya.

- **Region 7** (Company G) Headquarters is located in Abidjan, Cote d'Ivoire and is responsible for Western Africa.

- **Region 8** (Company H) is located in Frankfurt, Germany and is responsible for Central Europe.

- **Region 9** (Company I) is located in Fort Lauderdale, Florida and is responsible for the Northern Western Hemisphere.

Headquarters Company is composed of approximately one hundred Marines providing administrative, logistical, legal, training and educational support to the Marines around the globe.

Located within Headquarters Company, the mission of MSG School is to select, train, and screen Marines from any Military Occupational Specialty, male or female, for MSG duty. The School is eight weeks for detachment commanders and six weeks for sergeants and below. It is tough, intense, and grueling. While the 30-35% attrition rate is the highest of any Marine Corps school, the result is that attrition from the field is only 2-3%. This speaks volumes to the quality of the Marines serving on MSG duty.

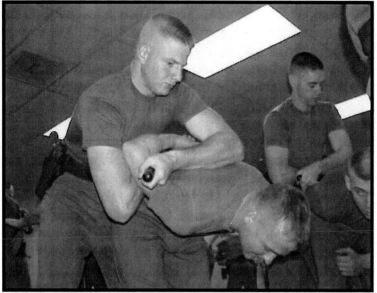

Students at Marine Security Guard School learn how to subdue a trespasser using a PR-24 side-handle baton.

Operations

MSG companies are commanded by a Marine lieutenant colonel. The company headquarters normally consists of two company-grade officers (first lieutenants or captains), a First Sergeant, an administration chief and two clerks. The company's mission is to ensure that administrative and training standards are maintained, as well as to advise the diplomatic posts in their region on the proper employment of MSGs. The company officers and First Sergeant visit each MSG detachment at least four times every year. They not only conduct formal inspections, but also observe detachment morale and meet with post officials. The results of the inspections are forwarded to the DOS.

At a diplomatic post, the Detachment Commander and

RSO form the Post Security Team. Their relationship is the key to the security program's success. The RSO is overall responsible for all internal and external security programs, as well as all background and criminal investigations. The Detachment Commander is ultimately responsible to the Chief of Mission, but normally reports to the RSO on day-to-day issues. At larger posts with several RSOs, the detachment may report to one of the Assistant RSOs.

The Detachment Commanders are responsible for the operation of the detachment and are considered "commanders" by the Marine Corps. This is a unique distinction because it is one of the very few times a Staff Non-Commissioned Officer (SNCO) can officially carry this title. Detachment Commanders can be staff sergeants (E-6) through master sergeants (E-8). Unlike the individual MSGs, the Detachment Commanders can be married. Detachment Commanders need not have had a previous tour as an MSG, though many have. Being a Detachment Commander is one of three "special duty" or "B-Billet" assignments that an enlisted Marine generally needs in order to have a successful career. The other two special duties are recruiting and drill instructor.

Detachment size is based on the individual diplomatic post's requirements. The local threat level, the size and layout of the building(s), and the amount of workday business are all taken into account when determining the number and frequency of MSG security posts. The minimum detachment size is one Detachment Commander and five MSGs (referred to as a "1 and 5" or "1/5"). This allows them to man one security post 24 hours per day, 365 days per year, while keeping the duty hours at a reasonable level so that the MSGs can conduct other routine training, internal management of the detachment, and have some time off.

Diplomatic posts that require more than one security post have proportionally more MSGs. Approximately 40% of the Battalion's detachments are 1/5, 40% are between 1/6 and 1/10, and the remaining 20% are larger than 1/10. The largest detachment is in Cairo with two SNCOs and twenty-eight MSGs. Detachments with more than seventeen MSGs are assigned two SNCOs - one is the Detachment Commander and the other the Assistant Detachment Commander.

"Post One" is the name of the primary interior security post. It is normally in the lobby or main entrance of the building housing the Chief of Mission. Post One is the principal command station for all access control to the building. It is equipped with fire alarm controls, closed circuit televisions, radios, and intrusion detection equipment. Residing behind bullet-proof glass, the MSGs survey the personnel traffic and monitor the various security displays around the clock.

At the larger diplomatic posts, additional security positions are labeled Post Two, Post Three, etc. They may be manned 24 hours per day every day, or just during normal business hours. They could also have a full complement of security displays and equipment similar to Post One, or they could be a roving security watch after the embassy has closed.

MSGs must always be prepared to conduct reaction drills, called "Reacts," to their embassy for emergencies such as fires, bomb threats, bombs, intruders, riots and demonstrations - and must be contactable via telephone, radio or pager at all times. Upon reaching the embassy they assemble in the "React Room" to receive orders and direction from the Detachment Commander. This room provides not only a storage area for weapons, ammunition, and personal protective equipment, but a safe and secure position to suit-up for the React situation. Each potential React scenario is practiced and has its own standardized drill from which the MSGs can modify to fit the actual situation.

In addition to the normal duties carried out at their home detachment, MSGs are often sent on temporary additional duty assignments to protect classified material during official Presidential, Vice Presidential, or Secretary of State visits overseas. Normally, MSGs man a security post in a hotel which the DOS and Secret Service Special Agents use as an operations center. MSGs wear civilian coat and tie while performing this duty.

LIFE AS AN MSG

MSG duty is considered a "good deal" by most Marines. They have an increased opportunity for promotion, live in conditions better than most Marines in the Fleet Marine Force, and get the opportunity to live in places they likely would never see if they weren't an MSG.

Since this is special duty there are certain restrictions placed upon applicants, for instance a recent policy dictates that Marines with certain tattoos may be excluded from embassy duty, citing the need to present a professional image abroad. Candidates are also required to be exceptionally mature and financially stable.

Marines intent on pursuing a position as an embassy security guard should also be aware that a good portion of the stint may be spent in a hardship post. The Corps categorizes duty stations in Tiers. Tier 1 is a city with amenities found in the U.S., such as London or Paris, Tier 2 is a city with some amenities found in the U.S., and a Tier 3 location has few if any of the amenities found at home. The intent is to assign a Marine serving in a Tier 3 country to a Tier 1 country for his follow-on tour.

After graduating from School an MSG can expect three one-year tours, while Detachment commanders will have two 18-month tours. MSGs can take Continuous Overseas Tour (COT) Leave between their tours, and while many return to the United States for their COT leave more than a few spend it traveling overseas.

Ultimately, MSGs and Detachment Commanders are assigned to a detachment based on the needs of the Battalion, however their preferences, any geographic restrictions, and input from the company commanders are factored into the assignment decision as well. Additionally, the Battalion uses a lineal quality of life ranking of all the detachments to ensure that an MSG who had a more austere first tour will get a "better" second tour and vice versa. Finally, some MSGs are further screened and nominated to serve at selected special duty diplomatic posts.

It is not uncommon for a Marine House to have a pool (lower right).

The majority of the Marines live in the civilian community in a house referred to as the "Marine House." Each gets their own bedroom and often their own bathroom, and the detachment manages its own mess fund to ensure adequate nutrition for all. Every Marine House is equipped with exercise facilities and areas to host social events, and some even have a swimming pool or large lawn for organized athletics or large gatherings. The MSGs also take advantage of the Armed Forces Radio and Television Service and the Naval Motion Picture Service to watch current shows and movies.

An MSG detachment is involved in every facet of life within the American community they serve. Whether hosting social events, sponsoring local community activities, or just generally adding to the rich experience of living overseas, our MSGs are indeed "Ambassadors in Blue." Recent events, however, have clearly highlighted the true benefit and mission of the MSGs. The unplanned and unexpected evacuation of the embassies in Freetown, Sierra Leone (May 1997) and Brazzaville, Congo (June 1997) were greatly facilitated by the actions of the detachments. Whether organizing convoys to the airport, destroying classified equipment, or providing protection to embassy personnel, the MSGs of these detachments where uniformly praised for their actions.

While these two evacuations are at the extreme end of the mission continuum, MSG detachments are called upon every day to react to their embassies for crowd demonstrations, fires, intruder alerts and bomb threats. Many an American citizen in trouble overseas has been calmed and reassured in the middle of the night when, in calling the embassy, they hear "This is Post 1, American Embassy, Sergeant Stryker, how may I help you?"

MSG DETACHMENT

There are, at any given time, about 140 MSG Detachments located around the world - and each one of them has its own personality. A Marine serving at any of those posts should understand the unique opportunity he has been given, and get everything out of the experience that he possibly can. It is a *tragedy* for someone to spend three or more years overseas in some exotic location and come back without having learned a language, visited famous landmarks, experienced a new culture, or any of a hundred other things (and it happens). On the other hand, it is not unusual for many other MSGs to return home with the *ultimate* souvenir - a spouse. Keep in mind, one of the requirements for sergeants and below on the program (Detachment Commanders are exempt) is they must agree to remain unmarried until their tour is over – but even so, a high percentage get hitched to a foreign national the day after their tour is complete.

What follows in this chapter is what I like to call "things I wish I had known before I went to my first post." While there is no substitute for experiencing things firsthand, it is also foolish for every single Marine who becomes an MSG to "reinvent the light bulb" when they come aboard. There are many little things the instructors can't, or won't, tell MSG students as part of the official MSG School curriculum – but I am not hampered by the same restraints they are. I think anyone going to post for the first time will find something of value in what I have to say, and hopefully those insights will help him or her avoid some of the stumbling blocks I encountered while I was out there. I also hope some of this information will help Marines in the fleet who are wrestling with making a decision about whether or not they want to volunteer for the program to make an informed choice.

There are three types of independent duties, or "B Billets," in the Marine Corps for enlisted Marines to choose from - Drill Instructor Duty, Recruiting Duty, and the Marine Security Guard Program. The first two are absolutely crucial to the future of the Corps. Every single Marine has come in contact with both a Recruiter and a Drill Instructor at the beginning of their enlistment, but MSGs are something altogether different. Remember, "Embassy Duty" did not exist prior to the 1940's, and if it ceased tomorrow the Corps would still go on.

Now, you may be wondering why I chose MSG duty over the other two if that's the way I feel, and it is a legitimate question - to which there is a simple answer. I didn't relish the idea of being a recruiter, a job in which I would have to more or less "suck up" to a bunch of fuzz-faced, snot-nosed high school kids in order to "make mission," and I was even *less* excited by the prospect of living the "Ground Hog Day" existence of a DI who has to go through the same training cycle with a different group of clueless recruits every couple of months. I realize there are great rewards which go along with each job, but they just weren't my cup of tea – and I tip my cover to all of the fine Marines who are now, or have ever been, out there doing those critical jobs.

On the positive side, I *chose* to become a Detachment Commander because I liked the idea of leading a group of hand-picked, highly trained Marines in the execution of a job which is important to the national security of our nation – and I'm glad I did!

Now for the "nuts and bolts." Most MSG detachments are

what is known as a "one and five" – i.e. one Staff NCO Detachment Commander, and five watchstanders. There are other, larger detachments to be sure, but I am going to focus on the 1/5 Det for illustrative purposes.

MSG Detachment Canberra, Australia at the Birthday Ball on 10 November, 1995. Left to Right are Cpl Evans, Cpl Soffe, GySgt Bufalo, Sgt Monnard, Sgt Franz, and Cpl Hendrix.

The first issues I am going to address are the ones which end the careers of many aspiring MSGs before they get started. That's because there are several criteria which must be met in order to be selected for training, but sometimes "fall through the cracks" during the screening process.

First and foremost, MSGs must be qualified to receive a TOP SECRET security clearance. That means there can be no skeletons in their closets – because believe me, the agents who do the SBI (Special Background Investigation) *will* find them. I know that firsthand, because one of the Marines in my student detachment was dropped *the day before graduation* because he omitted something on the questionnaire which would have disqualified him for a clearance. To make matters worse, he was one of the best students in the class - as well as the son of a former MSG Battalion Commander!

Another thing a prospective MSG must do is become computer literate. I realize that is becoming less of a problem in this day and age, since the kids of today seem to grow up with a pacifier in one hand and a computer mouse in the other - but there still a few holdouts out there.

Conversely, one of the problems which is NOT getting better as the years pass is the literacy of our young people. The written word is an important tool on the program because it is used in logbooks, naval messages, detachment orders, official correspondence and a host of other things. MSGs must know how to spell (or at least how to use spell check), but just as importantly they have to use proper grammar and punctuation. I for one made it a point to allow my troops to draft their own message traffic, which I then red-penned and sent back for correction – and it usually took four or five tries before I would give them a thumbs-up. It wasn't long before they learned to do it right the first time, and that is a skill they were able to take with them when they moved on to bigger and better things in life. The bottom line is, nothing says "amateur" or "dummy" louder than a logbook entry that looks like it was made in crayon by a four-year old - so take the time to learn!

There are a lot of important duties for MSGs to perform, but at the top of the list is pulling duty on "Post One." It is so vital, in fact, that the official MSG Battalion newsletter is named *Post One*. State Department employees simply see a Marine "standing in a box" behind a plate of bulletproof glass when they come in every morning, and then again when they leave for the day, and think his only job is to buzz them through the hardline door. They couldn't be more wrong. The Marine on duty at Post One is the embassy's first line of defense, the focal point in a crisis, and the "face" of the embassy (and the United States) in the eyes of visitors. When a foreign national, or an American traveling abroad for that matter, visits a U.S. diplomatic mission overseas the first person they usually see is that MSG on Post One – and he had *better* be squared away!

Post One is responsible for determining who will and who will not be granted access to the mission (within certain parameters), monitoring close-circuit feeds from various parts of the compound, answering the embassy phone lines, controlling radio nets – and a whole lot more. The Marine manning that post has to be on his toes at all times! Sometimes in larger embassies there are other posts to stand, including "roving" posts, but Post One is still the boss.

The Marine manning Post One is the first American a visitor to an Embassy meets, so he must be squared away at all times.

One of the most important, and most mundane, duties is making logbook entries. The Post One Log is the official record of the embassy, and it is imperative for it to be accurate. *Everything* goes in there. My motto was, "when in doubt, log it in." Even so, some Marines get a little behind the power curve sometimes with the intention of "catching up" when things quiet down - but that is a big no-no. The easiest way for the Detachment Commander and RSO to nip that sort of practice in the bud is for them to drop by Post One periodically (which they should be doing anyway) and reconciling what has been going on with what is in the book.

Another boring but vital duty performed by MSGs is called a "Char Escort." That entails escorting and closely supervising the cleaning crew (the "chars") who come into the embassy to empty trash cans and sweep the floors each night. The chars are almost always foreign nationals, and must be watched 100% of the time. Not 99%, but 100%! If the Marine on duty were to leave them alone for a minute to make a head call, or just looks away for a moment when the phone rings, there is no telling what might happen. It would only take a few seconds for a char to pick up an unsecured classified document, or plant a listening device – or a bomb, for that matter!

I just mentioned the possibility of there being unsecured classified materials lying around, but that should not be the case if the detachment is doing its job properly. A classified material "sweep" should be conducted each night before chars, because State Department types are notorious for leaving things "adrift." In fact, it's *unbelievable* how lax security can be at some embassies. Sometimes safes are left unlocked (or they are just not "spun off"), passwords are written down in supposedly "secret" locations, and secure computer terminals are left logged on – you name it, they do it. When security lapses like those are found, it is the Marines' job to issue a "violation" slip and secure the material immediately – and here is where the fun begins. The State (or DOD) employee who transgressed will likely challenge the validity of the "bust" the next morning, because it will put a black mark of their record if the RSO concurs with the MSG.

Some detachments have made an effort in the past to tighten security at their post by making something of a "contest" out of the issuance of violation, with the Marine who writes the most in a given month getting some sort of "prize" – such as an extra day off. On the surface this may seem like a good idea, but it really is not. The State Department workers will get wind of what is going on eventually, and will accuse the Marines of "headhunting" and manufacturing bogus violations.

At the other end of the spectrum is the lax detachment with an easygoing Detachment Commander who wants to avoid friction with their counterparts in the Department of State. They drop hints, make exceptions, and turn a blind eye to all but the most serious violations. That is a recipe for disaster. The best thing to do is remember the old adage - "A Marine on duty has no friends."

One of my biggest mistakes as Detachment Commander was not standing enough duty on Post One. Like most SNCOs I did it during Post Familiarization when I first arrived in-country, and filled in when we were shorthanded for one reason or another – but in retrospect I would have had a much better understanding of what goes on behind the scenes if I had pulled a shift at least once a month.

That's not to say there isn't a lot for a Detachment Commander to do in the normal course of his job – if he does it correctly. I say that because some SNCOs get a bit too laid back when they get away from the FMF and don't keep their "nose to the grindstone" as much as they should. This is particularly true at the nicer posts where there are a lot of distractions and recreational activities to choose from. I found out about *that* the hard way. When I arrived in Australia I made the mistake of not going over everything with a fine-toothed comb before assuming command, and it ended up costing me dearly. Our Operations Officer flew in from Company Headquarters in Bangkok to conduct a regularly scheduled SAI (Semi-Annual Inspection) two weeks after my arrival, and he found discrepancy after discrepancy. Keep in mind that there is no way I could have corrected all of the problems even if I had *known* about them (it ended up taking six months), but when you are in command you are responsible – period. It turned out that my predecessor (a very nice guy, by the way) had spent a disproportionate amount of his time "Down Under" golfing, sightseeing, and learning to fly – and had neglected his duties (primarily paperwork). Don't make the same mistake.

Another job which falls to the "Gunny," as most State personnel refer to the Detachment Commander (regardless of his actual rank), is conducting after hours checks at the Embassy and Marine House. This is one of the most important tools a Detachment Commander has to keep his Marines in line, and it is also the one they make the most mistakes in implementing. The idea behind the checks is to show up, unexpectedly and unannounced, in order to see if the "mice are playing" while the "cat is away." It is common for Detachment Commanders to tip their hand by doing things such as writing it on their calendar, saying "I'll see you later" to the Marine on duty at the end of the work day, alerting the A/Slash ahead of time, always coming in on the same day of the week at the same time - or not coming at all.

Many SNCOs have been relieved of their commands because they were too lazy to roll out of the rack at 3 AM and drive to the embassy (I can still hear my ex-wife saying, "Do you *have* to get up now?"), so they instead called Post One and told the Marine on duty to make a (false) log entry saying they were there. My advice to the Marine on Post One who is faced with such a situation would be to say, "Gunny, please don't put me in a position where I have to lie to the Inspecting Officer during my next SAI interview." My advice to the "Gunny" would be - do your job!

While I am on the subject of checks I must stress the importance of making the rounds at the Marine House as well. The most common reason for sergeants and below to be removed from the MSG program is unauthorized civilians in the BEQ (and in their rooms). Most Detachment Commanders don't *want* to catch their Marines violating this particular policy, and subconsciously (and sometimes purposely) make it easy for their troops to "get over." They know that if they always make their BEQ visits on the same nights as, and immediately after, their check at the embassy, the Marine on Post One will call his buddies with a warning the moment he leaves the building. I maintained order (to the best of my knowledge) by making my A/Slash (who was the senior Marine in the BEQ) personally responsible for any infractions, and relying on him to keep an eye on things when I was not around. I just cannot stress this strongly enough - keep in mind that *entire detachments* have been sent home because of such practices - so decide for yourself.

This is probably an appropriate juncture to talk about Liberty and related matters. The opportunity to walk the streets of Paris, drink ale in a London pub, or go on safari in the African Serengeti is one of the biggest benefits of MSG duty - and it is also one of the biggest hazards.

MSGs who take advantage of the unique liberty opportunities at post can see places most people only read about.

The upside is obvious - a once in a lifetime opportunity to travel the world, see amazing sights and experience other cultures - but the negative is always lurking. Young Marines in a foreign land with the keys to an embassy are a tempting target for foreign intelligence agencies. One only needs to review the involvement of Clayton Lonetree in the Moscow Station scandal to know that is true. But even in the absence of cloak and dagger escapades, there are other potential pitfalls. A liberty incident in the United States - whether it be a bar fight, a car crash or a pregnant girlfriend - is unfortunate, but not the end of the world. Those very same transgressions in a foreign country could easily become *international incidents* and embarrass both the Marine Corps and the Nation. That's why the screening process for MSGs is so stringent and the leadership responsibilities are so great.

One of the best tips I can give a Detachment Commander is get to know your Marines, as well as the people they are socializing with. Don't hole up in your quarters while off duty, but at the same time you shouldn't act like "one of the boys" either. It's a fine line to be sure - but that's "leadership" in a nutshell! It is also important to make local contacts, because knowing the "right people" both inside the embassy and out in the local community can often be the key to getting things done.

One of the best ways to integrate a detachment into the international community is through participation in Embassy social functions. While most Marines are relatively young and some can be a bit rough around the edges from service in the FMF, there is no getting around the requirements of the diplomatic world. That doesn't necessarily mean big strapping Marines with recent combat tours under their belt need to sip tea with their little fingers in the air, but it does mean they must "show the flag" at appropriate events – and conduct themselves accordingly while doing so. That goes double for the Detachment Commander.

Not surprisingly one of the biggest annual affairs at American diplomatic missions is the annual Fourth of July celebration, but that is just the tip of the iceberg. The type and frequency of events of course differs from post to post, and it is primarily driven by the customs of the host nation as well as the local security situation. Parties may be the order of the day in places like London and Paris, but are a rare thing indeed in Mogadishu and Baghdad! Often Marines are simply encouraged to attend, but sometimes detachments are asked to supply a color guard or provide other support - so it is a good idea to organize and practice these things regularly because you never know when something might be scheduled on short notice.

While the diplomatic community will certainly provide many opportunities to mingle with embassy staff, foreign diplomats and even the local population, it is also a good idea for the detachment to organize and host periodic social events in order to foster goodwill. While the Marine Corps Ball is without question the centerpiece of any detachment's outreach program, it is more of a showpiece than a networking opportunity because of all the pomp and circumstance involved. Monthly or weekly Marine House events such as movie nights, chili cook-offs or pool tournaments are common, but sometimes a formal occasion is called for. I found Mess Nights to be the perfect way to develop camaraderie while observing the traditions of the Corps, and made it a point to hold one per year during my four years as a Detachment Commander. It takes a bit of work to put one on to be sure – but it is well worth it!

Fourth of July Mess Night at the Royal Military College Sergeants' Mess in Canberra. The guest of honor, Colonel Stephen Saunders of the British High Commission, sits to the author's right. He was assassinated by terrorists while serving in Athens several years later.

39

There is one more dimension, aside from the operational and social aspects, to MSG duty. The "grunt work" of being a Detachment Commander is the paperwork. Detachment Orders must be written, financial ledgers must be balanced, and message traffic must be drafted. I guess that's why Admin Chiefs tend to make such good Det Commanders (at least until the shooting starts) – but no matter what your background, before doing anything, it is imperative that you carefully read and re-read every detachment order - even if you are a "second poster."

The paperwork is a grind to be sure, but if you are lucky your A/Slash will be the locked on and squared away sort who can be trusted to take part of the load - in contrast to the hard charger type who is great in a gun battle, but can't count to eleven without taking his shoes off. I have had both.

Assistant Detachment Commander

A good Assistant Detachment Commander is worth his weight in gold. As the second-in-command he must be prepared to take over when the Gunny is away (and should be trained accordingly), and is responsible for a variety of administrative functions such as preparing the duty rosters. The true test comes when a Detachment Commander can go away for a few days and not feel compelled to call every five minutes to check on things (such as when attending the annual Company Conference out-of-country). In a lot of detachments the A/Slash also plays the "heavy" disciplinarian to the Detachment Commander's "wise father figure," because a lot of SNCOs feel their Marines will be more likely to confide in them when problems crop up if they are perceived as being more accessible.

The position of A/Slash is not the only "collateral duty" assigned to MSGs - not by a longshot. Additional duties are assigned to each and every Marine in a 1/5 detachment. The Training NCO is responsible for off-duty education in addition to operational training, the Mess NCO makes sure the Marines are properly fed, the Supply NCO handles logistics, and the Bar NCO runs detachment functions and social events. These duties should be rotated periodically so that every Marine in the detachment gets a chance to try their hand at each, although some detachments are hesitant to do this for fear of mistakes cropping up as a new Marine gets "snapped in."

Training NCO

Regular re-familiarization with TO weapons is crucial to the MSG missions, although finding a suitable range can be difficult in some countries.

The job of the Training NCO is arguably the most important, because it impacts directly upon the operational readiness of the detachment. Each week at "Guard School" (an informal meeting of the MSGs) operational issues are discussed and refresher training is administered - and it is the Training NCO who is responsible for making it happen. In a lot of detachment "training" tends to be limited to taking MCI courses, weapons requalification, and an occasional PT session, but there is so much more that can, and should, be accomplished. Physical Training is an example of an area where most detachments can improve. Detachment Commanders who limit PT to a weekly three-mile run with a few pull-ups and crunches thrown in are missing a great opportunity to challenge their Marines and improve motivation. The "PFT disciplines" are important to be sure, but after awhile such a regimen becomes boring and repetitious – so organized athletics, competitive events and operationally oriented training should also be included periodically. For example, my detachment played rugby against diplomats from the French Embassy in the Congo, and in Australia we formed a fast-pitch softball team (supplemented by embassy personnel) as part of a local league. Another great way to work some cardiovascular training into your program is through participation in local road races. While in Australia I ran everything from a 10K to a marathon, and many of my Marines did the same. Liberty and PT can even be combined by participating in the local "Hash House Harriers" club, which has chapters all over the world. For more information on that organization, go on Amazon and pick up a copy of the book *Beer Soup For the Hasher's Soles.*

The Training NCO is also responsible for coordinating a wide variety of training evolutions, from first aid refreshers to language training. The difference between a good detachment and a great one often hinges on his or ability to think out of the box and come up with a program which is both enjoyable and beneficial.

Mess NCO

The Mess NCO is responsible for feeding the detachment, but that doesn't mean he or she slaves away over a hot stove. Duties include planning the menu, shopping for food, hiring, supervising, and paying the cook (if the detachment has one), and keeping track of the mess fund. Math skills are important, because the each month the Mess NCO is required to collect the BAS (Basic Allowance for Sustenance, aka "ComRats") of each watchstander and deposit those funds into a special account. A meticulous ledger and vouchers of all transactions must also be kept, and periodic audits are conducted to ensure the mess fund is balanced to the penny.

Supply NCO

The Supply NCO is, in effect, the "S-4" or logistician of an MSG detachment. This Marine is responsible for ordering, inventorying and maintaining operational and recreational equipment as well as replenishing perishable and/or expendable supplies. Most of this is done by composing message traffic to Battalion Supply, although some items not in the MSG Battalion supply system are procured through the Embassy's General Service Office (GSO). Due to the high volume of messages which must be drafted, I highly recommend taking courses in drafting Naval Messages and spelling.

Bar NCO

All MSG Detachments are required to have a Bar Fund, which some say is quite appropriate given that the Corps was founded at Tun Tavern. The primary purpose of the Fund is to support the financing of the annual Marine Corps Birthday

Ball, although the purchase of recreational items not available through the supply system or State Department may also be authorized by the Detachment Commander when appropriate. Like the Mess NCO, the Bar Fund NCO must maintain meticulous records and a bank account which are subject to periodic inspection and audit - so once again, good math skills are required.

A typical Marine House bar area. In some "dry" countries, the detachment watering hole is the only place to get a drink.

Monies are raised through the sale of Detachment T-shirts and similar novelty items, as well as the conduct of periodic social functions at the Marine House or other appropriate venue at which refreshments are sold.

One of the biggest responsibilities of the Bar NCO is assisting the Detachment Commander with the planning, financing and execution of the Marine Corps Ball. This is considered the premier social event of the year at most embassies and consulates, and since it is a direct reflection on both the local detachment and Marine Corps as a whole it must be carried out flawlessly. Planning should begin many months in advance, addressing such details as choice of venue, menu, seating charts, programs, ceremonial music - and of course, the birthday cake.

The author cutting the cake during the 223rd Marine Corps Birthday celebration in Canberra, Australia.

State Department

I have focused on the Marines up to this point, but it is also important to mention the State Department. It is a fact of life that while the military in general, and Marine Corps in particular, tend to be relatively conservative in nature, the Department of State's Foreign Service Officers tend to be liberal – and this can sometimes generate a bit of friction. Don't let these dueling ideologies get in the way!

One of the biggest challenges for a Detachment

Commander, and all MSGs for that matter, is dealing with dual chains of command. It is almost impossible to serve two masters, but that is exactly what the Memorandum of Agreement between the Marine Corps and Department of State requires. A DetCmdr must answer to a Company Commander located far away in another country (unless he is "lucky" enough to command the Det where Company Headquarters is located) who makes short semi-annual visits, and a Battalion Commander all the way back in Quantico he never sees. That is doable because all are Marines, and everyone is for the most part on the same sheet of music. Contrast that with the State Department chain of command, which consists of the Ambassador/COM, Deputy Chief of Mission, Admin Officer and Regional Security Officer. The DetCmdr sees them every day, and for the most part they have no clue about Marine Corps regulations, traditions or mindset. If the Ambassador wakes up one morning and decides he's like the Marine on Post One to wear pink in observance of breast cancer awareness, it is the Detachment Commander's job to tactfully reconcile Marine Corps uniform regulations with the COM's desire. Depending upon the personalities involved, this can be one of the most challenging aspects of being a Detachment Commander.

Since we are discussing things relative to the Chief of Mission, let's talk about the different types. There are two flavors of Ambassador – the political appointee, and the career diplomat. Political appointees are most often found at high profile posts such as London and Tokyo and usually get the job because they are friends with, or raised a lot of money for, whoever is occupying the White House at the moment. Career diplomats, on the other hand, are professional diplomats who have come up through the State Department ranks. Since they have usually served in several different billets at numerous posts before receiving an Ambassadorship these individuals are usually far more familiar with the functions of the Marine Detachment and therefore much easier to work with - although that is not always the case. It should also be pointed out that not all Chiefs of Mission are Ambassadors, which is important to know if you are posted to a Consulate (which is subordinate to an embassy, and usually located in a major city other than the capital) or a Liaison Office (located where the United States does not have formal diplomatic relations).

The Deputy Chief of Mission (DCM) is second in the pecking order, as the name implies. At posts with politically appointed Ambassadors the DCM, as a career diplomat, is the person who really runs things – while the Ambassador is more or less a figurehead. Since the DCM runs day to day operations, it is important for the detachment to have a good working relationship with him/her.

After the DCM comes the Administrative Officer. Most of the time there is not a lot of contact with this officer, although there will be times when the DCM is away during which this person will move up and fill that billet, and on other occasions he/she may serve as an acting RSO. Aside from that, the Admin Officer holds the purse strings, and will often be the person who determines financial support for the certain detachment needs such as housing.

The final and most important member of the State Department hierarchy - at least from a Marine's perspective - is the Regional Security Officer (RSO). The RSO is a representative of the Diplomatic Security Service (DSS), and as such is the person with direct operational control over the detachment. As a result, it is imperative for the detachment to develop a good working relationship with this person - but as you will see later in this book, that isn't always possible.

47

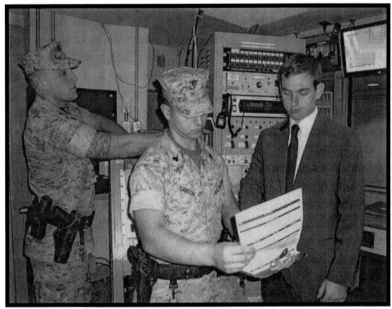

All of the paperwork and politics at post are fine and dandy, but the focus of the detachment and RSO must be operational readiness.

Operations

When it comes right down to it, all of the diplomatic and ceremonial functions of an MSG detachment are secondary to its operational responsibilities. The Marines are responsible for the security of the mission and safety of its personnel, *and if they do nothing else* they must be prepared to meet every threat from intruders to angry mobs.

In order to do these missions effectively, periodic and unannounced react drills must be conducted to ensure the detachment's operational readiness is honed to a fine edge. These scenarios include dealing with telephoned bomb threats or suspicious packages, locating and apprehending intruders inside the chancery, and dealing with demonstrations or mobs at the gates or on the perimeter. Detailed operational plans must be developed by the Detachment Commander and RSO, and should be modified from time to time as facilities or circumstances change. The key is to be creative, and not take a "cookie cutter" approach when moving from post to post.

In order to carry out its mission a detachment must maintain its equipment in a perpetual state of readiness. Weapons and ammunition, radios and call signs, and individual react gear are just a few of the things which must be ready to go on a moment's notice. MSGs can never count on the "cavalry" coming to the rescue, because shifting political winds in any country can mean a change in the support and security to be expected from a host nation – and although there have been many instances in which the U.S. has dispatched military forces to shore up the defenses of a vulnerable diplomatic mission, that is not always possible due to the realities of geography or politics.

While the RSO is charged with the day to day supervision of the detachment's readiness, the Marine Corps' chain of command also takes an active role. Twice per year Semi-Annual Inspections (SAI) are conducted by an officer from company headquarters, usually the executive or operations officer, and while they do go through administrative items such as training records and financial audits with a fine toothed comb the most important evolution is a demonstration react drill during which he can assess the detachment's operational readiness. Additional drills are also run during the Company Commander's periodic Command Visits, so it pays to be ready at all times.

Another key operational responsibility is the security of classified material. That can mean its destruction during the imminent breech of a secure area, or routine sweeps of office spaces for unsecured classified documents. This tasking in one of the primary reasons MSGs are required to possess a Top Secret security clearance.

You never know who might drop by. Here the Chairman of the Joint Chiefs visits with MSG Detachment Bogota, Columbia.

Assorted Tidbits

The preceding contained some of the basic information a Marine should know in order to have a successful tour on MSG Duty, but there are a thousand other tidbits which will come to light only through experience because no two posts are the same. When it comes time for a new MSG to put in a "dream sheet" for his or her first assignment, a lot hinges on what your priorities are when making a selection (although the odds are good you will go somewhere else). For instance, some high-profile posts have frequent VIP visits, and others are located in quiet third world backwaters where nobody goes. Some MSG detachments allow Marines to drive, while others require the hiring of an FSN as a driver. Some posts like Tokyo have extremely high COLA (Cost of Living Allowance) which potentially allows the saving of a great deal of money, while others don't. And so on. The point is, do the research and find out what you may be getting yourself into.

One quick story which serves as a microcosm of the MSG experience concerns the type of barber used at each of my posts – which is appropriate, given the nature of Marine Corps grooming standards. In Brazzaville it was necessary to have our driver bring us to a little hut in the middle of the Bacongo district which had no electricity or running water and which we could never have located on our own, so that a little fellow with a dull, rusty razor blade could give us the closest thing to a regulation haircut we could manage given the language barrier – and our fear of contracting tetanus. Contrast that with Canberra, where a professional, English speaking gentleman named Steffan has been cutting Marines' hair for decades in a clean, modern shop and knows more about our requirements than most of the barbers at Camp Lejeune. It's a big world out there, and different experiences like that are what make MSG duty so interesting!

GEOGRAPHY LESSON

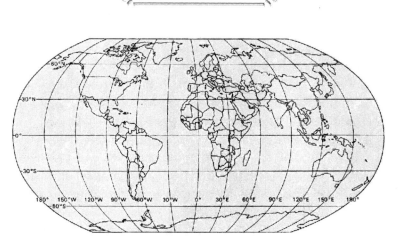

How many Americans have ever heard of places like Ouagadougou, Burkina Faso and Quito, Ecuador – much less be able to find them on a map? Maybe one in a hundred... and the odds are pretty good that *one* is a Marine Security Guard! That's because waiting for that first assignment after MSG School is like playing a game of Russian Roulette – where you could end up almost *anywhere* on the globe. Then, fifteen months later, it's time for the 'second post' assignment, and every MSG due to move keeps checking the message traffic to see which posts will have openings. You just never know.

While a few Marines just don't care and are happy to go wherever they are sent, most spend a lot of time doing their "homework" – going through post reports in order to get a better idea of where they would like to be assigned next. Each MSG is required to submit a "wish list," and although there are no guarantees, every effort is made to accommodate them based upon manpower requirement and the battalion "lineal list."

The lineal list ranks each detachment based upon a combination of location, hardships, housing and a number of other factors, with so-called "luxury" posts like Canberra and London ranked near the top, and "hardship" posts like Mogadishu and Kabul near the bottom. In order to be "fair" (or as close to it as you can get in the Marine Corps) the battalion tries to assign each Marine to one "good" post and one "bad" one – but it doesn't always work out that way.

The concept of good vs. bad postings is really subjective in nature. While one MSG may want to explore all of the cultural sights Europe has to offer, another may be just as happy in the backward and often dangerous (though often high COLA) countries of Africa.

The bottom line is there are a lot of opportunities on the MSG program, and what often seem to be miserable assignments at first often turn out to be the most rewarding. Any way you look at it, it's a *heck* of a way to learn world geography!

IN EVERY CLIME & PLACE

THE VERY FIRST
Embassy Marines

MSG Detachment Paris

Harold Stephens

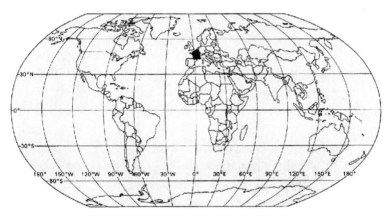

I marvel every time I see U.S. Marines guarding American embassies around the world. My interest in the Marine guards, however, is more than casual. Let me tell you why.

I was a Marine sergeant at the Naval Air Technical Training Center near Memphis, Tennessee when my company commander called me into his office. But what had I done wrong? This looked serious. He dismissed everyone in the office, his orderly and secretaries. I stood at attention, nervous as a recruit in boot camp on Parris Island. But it was not Parris Island - it was NATTS. The date was September, 1948. The CO ordered me to be at ease and then picked up a brown envelope from his desk. "These are secret orders," he

57

said, "and you have to swear to secrecy. Do you agree?"

Agree to what? He read on. Marine Corps Headquarters at the Pentagon was selecting two hundred Marines from posts around the world. Each Commanding Officer had to select a man best qualified for the job. And what was the job? The CO didn't exactly know, except it was a request from the State Department and classified secret. He surmised it had to be some kind of cloak-and-dagger work. It sounded more exciting than ever. I accepted. I would be transferred to Arlington Hall in Virginia outside of Washington, D.C. How I wanted to tell everyone about the assignment... but I couldn't. I couldn't tell anyone. Not my buddies, not my family, not the young lady I had just proposed to. When I told her I was leaving, and couldn't tell her why, she didn't believe me. I got a slap in the face and that ended the engagement.

In Washington we assembled at the State Department. We were told, finally, that we were selected to guard the U.S. Embassies around the world. It appeared there were leaks at the embassies and in addition to guard duty we were to keep an eye on internal security at the embassies. This meant we had to be trained, which called for instructions from the FBI, CIA, and a couple other snoop outfits. We were ushered to a big clothing store in downtown Washington and issued civilian clothing, everything from double-breasted suits to underwear, shoes and socks. Some of us looked like real dandies, having forgotten how to wear stylish civilian clothes. Nearly every one of us were combat Marines from the Pacific war. Imagine some guys outfitting themselves with suspenders, bowties and black-and-white shoes.

I got my assignment, along with four other Marines, at the U.S. Embassy in Bangkok, Siam. The name Siam hadn't been changed yet to Thailand. I don't think there was a happier Marine in the Corps than I was. There was a song everyone was singing at the time called *Going to China* - or maybe it was Siam.

I had just served in China not long before, but the Marines were ordered to leave the country as Mao Tse-tung and his communist forces began to overrun the country. Now I was going back. How exciting. But then came the shock. Two days before I was to leave I was pulled from the assignment and told I was being court-martialed. In my security clearance I had stated that I had never been arrested. I had been, but it was nothing - not even worth reporting. I was once arrested in a barroom brawl and the judge, who had lost his legs on Okinawa, dismissed the charges. I thought that was the end. No more said. The judge told me to go back to the barracks and forget it, and we talked a bit about Okinawa. I was lucky. The reviewing officer at Marine headquarters had also been an infantry officer on Okinawa, so the charge against me was dropped and I was re-assigned to the U.S. Embassy in Paris, France. I didn't want to go to France - I wanted to go to Asia - but once I got to Paris I changed my mind.

In January of 1949 travel was still by ship. I had returned from China by air, one of the very first to do so, and it took five days, stopping at islands en route to refuel. Thus, we had to travel by ship to Europe, and the weather beaten old Gunny in charge of our detachment made sure it would not be commercial but aboard a Navy ship. It meant we had to travel by train to Portsmouth, New Hampshire to the Naval base and sail from there. Someone had failed to figure out that since we would be traveling as civilians, we would need luggage. "No need," the Gunny said. "That's why we were issued locker boxes." We had to label them with big letters PARIS, FRANCE. We went aboard the Pullman, still under secret orders not to tell where we were going. How embarrassing when passengers asked us where we were going and we couldn't tell them, or we'd be court-martialed if we did, despite the nametags hanging down from the locker boxes.

The Gunny found a destroyer sailing to Le Havre, France. He got us aboard, bunking us on canvas cots in the galley. At last, after a rough crossing, we were nearing Le Havre when the destroyer got orders to join naval maneuvers in the Mediterranean and sailed past Gibraltar to the Med. Then the State Department intervened, and we were order to proceed immediately to Paris. Easier said than done. In the mid-Mediterranean we twelve Marines were transferred, with our locker boxes and their tags, to a small destroyer escort and taken to Gibraltar. I had always wanted to see Gibraltar after reading Richard Halliburton as a kid, but this was not quite the same. Unfortunately Gibraltar was British, and the British were having a feud with Spain. We couldn't enter Spain, and were stranded on the rock, but we could fly to Tangier in North Africa - which we did. After a week, still not telling people where we were going, we flew to Paris.

I don't know how it is today, but in 1949 we were given a living allowance and had to find our own accommodation. Our CO, Lieutenant Breckenridge, had gotten us a good deal at the Continental Hotel on Rue Rivoli, one of the smartest hotels in Paris. (Today a room at the Continental is US$400 a day.) The bellboys toted our heavy locker boxes to our rooms, and we settled in. No one, for the first time, asked us where we were going.

And so began our tour in Paris. I have passed through Paris several times in the last few years and stopped to take a look at the embassy. I couldn't even get near the place. The well-fortified embassy today is nothing like it was sixty years ago. People could come and go freely then. I was rather fortunate, as I was assigned as orderly and bodyguard to Jefferson Caffery, the U.S. Ambassador. For my first year in Paris I traveled wherever the Ambassador went. Once we were in Cannes in southern France, and he wanted to visit the casino at the Carlton. I was of course armed with a .45 in a shoulder holster, and as far as I know am still the only armed person ever to be allowed to enter the casino.

The Champs Elysee in Paris, France circa 1949.

Thus when I see Marine guards today, my mind goes back to those years. After my discharge I turned to writing, and one book I wrote was about Marines in China called *Take China, the Last of the China Marines*, and another I had to write as a novel on my experiences in Paris was *The Tower & the River*. In *The Tower* I was able to write about Paris back in the days when Ernest Hemingway hung out in

Harry's New York Bar, and he would greet us with a salute, or we could chat with Orson Wells at the Night & Day on the Champs Elysees. We often stopped at Night & Day after work, and Orson was there after his stage performance. Orson liked the Marines, and always greeted us – but of course the whole world was different then. Embassy duty too may be different today, but I am sure the Marine Corps spirit is still there.

ABOUT THE MARINES

MSG Detachment Unknown

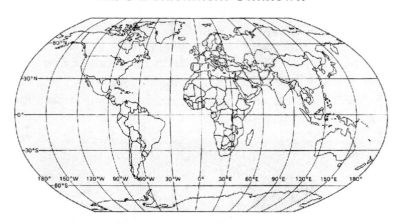

Many State Department employees assigned to our diplomatic missions abroad tend to marginalize Marine Security Guards and subconsciously (or sometimes overtly) classify them as "second class citizens." It's the old story of not noticing a policeman until a crime takes place and you suddenly NEED one. Fortunately, there are a few in DOS who appreciate all that MSGs do:

Maybe it's the Christmas season. Maybe I'm just getting old. Maybe I've been working overseas in some pretty mean places just too long and would rather be driving a turbo-charged American muscle car across the Nevada desert (we all have different fantasies, right?) Maybe it's the constant news crawl on my TV set announcing another dead American in Iraq or Afghanistan. I don't know, but for the last few days I can't stop thinking about how much we owe

63

the young Marines who protect our Embassy and how angry I get every time I hear that a Marine has died in Iraq. And today was particularly bad as CNN reports that seven Marines died in clashes against terrorists.

I don't want to denigrate any of our other fine armed services, but at State we have had a long and special relationship with the USMC. Since 1948, some of the best Marines get seconded to us to protect our diplomatic missions abroad. In addition, of course, both before and since 1948, it's the Marines who come yank us out when it all goes to hell or, as in Somalia and Liberia, or to save the Embassy from a howling mob.

Post One at American Embassy Kabul, Afghanistan.

In our Embassy in this rather tough corner of the Far Abroad, we get daily threats of all types - almost daily demonstrations in front of the fort-like Chancery. We shuttle about in armored cars, and have weekly "duck-and-cover" drills. We've had some nasty and very lethal bombings, and know that the bad guys are out there. They have us under surveillance, and they have lots of time, explosives, and guns.

We also have a detachment of MSGs (Marine Security Guards), all of them very young (18-23) led by a quiet but tough "old" Staff Sergeant (I doubt he's thirty) tasked with protecting the Chancery building and ensuring that we follow good security practices (note: one of the most dreaded events in the Foreign Service is to walk into your office and find a "pink slip" on your desk left by an MSG who the previous night found a classified cable left in an outbox, a safe not properly spun shut, or some piece of classified gear left unsecured. Those "pink slips" are career killers; of course, in the old Soviet GRU, one of these "security violations" was literally a killer... it meant the death penalty.)

Most days, however, you're hardly aware that an MSG is there: Just a shadowy figure standing inside a glass box, buzzing you through the hard line. Normally you sweep past him (and increasingly her) absorbed in your own thoughts, blabbing away on your cell phone, adjusting your tie, fumbling with papers, or just plain too rude and self-important to say "Good morning." When you have events at your house you rarely think of inviting the Marines. But despite all that, they remain cheerful, upbeat and exceedingly polite, and exude a quiet confidence that comes from great training and dedication.

Among the MSGs at this post we have two fresh from combat in Iraq, and itching to go back. These youngsters, one nineteen, the other twenty-one (both younger than my kids!), seem genuinely puzzled when we civilians ask, "So what was it like?" They can't seem to believe that anybody would be interested in, much less amazed by hearing about coming under mortar attack or driving a truck at high speed down some "Hogan's Alley-type" street lined with crazed and armed Jihadists. They relate it in a shy, matter-of-fact manner, full of military jargon. And they want to go there, again.

Watching these guys as they pulled toys out of the big "Marines' Toys for Tots" box in the Embassy lobby and hearing their cheerful shouts of "Oh, cool! Check this one out!" I couldn't help but think, "They're kids. They're just kids. Probably not much older than the orphans to whom they'll give those toys." I kept thinking about my own kids, living safely in the States, and the fact that they're older than these kids, these Marines.

But then I went with the "kids" out to the gun range. Suddenly they became deadly serious. The "kids" disappear. No goofing around. Strict discipline and concern for safety kicks in. They certainly know firearms, and treat them with respect and care. It was quite a sight to see the former "kids" deliberately, methodically pumping out rounds from their M-4s - single shot, three-shot bursts, full auto - punching out quarter-size groups in targets I can barely see. They don't look like kids anymore. They look like Hollywood's idea of Marines, like the actors John Wayne "led" in *Sands of Iwo Jima*. Now my thinking shifts to, "I wouldn't want to go up against these guys." And for a brief, very brief moment, I almost feel pity for the poor stupid thugs in Fallujah who had dared tangle with the Marines, "You jerks haven't got a chance. Just call Dr. Kevorkian and get it over with."

We all have had our days when we rant and rail against America's youth. I have heard my father's voice emanating from my own mouth: hopeless, hedonistic, rock addled, etc. I take it all back. I don't know what the Corps does to those orange-haired kids I see hanging out in the malls when I go home to the States, but whatever it is, keep doing it. The Europeans and their imitators in Ottawa, New York, Boston, and Hollywood paint their faces white and prance around in the "theater of the street" calling for peace. They wave their oh-so clever "Bushitler" posters, and over their lattes they decry the primitive "Red State" Americans. I know it's way too much to ask such smart and sophisticated people, but maybe they should take a moment to remember that it's these kids, these Marines from small-town America who put their own lives on the line to make all that noise and color of freedom possible. These kids, these Marines are the wall holding back the fascists of this century, and keeping the rest of us free.

Life isn't fair. The odds are not even. But I don't think these Marines would have it any other way. Semper Fi.

SALT AND PEPPER

MSG School, Quantico VA

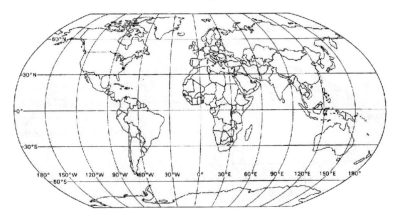

MSG training is more arduous for some Marines than others, often because those selected for the program come from a wide variety of Military Occupational Specialties. Marines with an infantry background generally have an easier time with physical training, those from Admin MOSs do well with the paperwork, and MPs are ahead of the power curve during law enforcement-type training – but there are some things you just can't "prepare" for:

One of the things I really enjoyed about MSG School was physical training. That's because while it *was* demanding for a lot of Marines, it was a piece of cake for me. My previous assignment had been with a Force Recon Company, and while I had managed to hold my own while there, I was hardly a "PT Stud" like some of the other guys. MSG School gave me a chance to run at the head of the pack for a change!

MSG students undergoing PR-24 side handle baton training.

Since I was a Gunnery Sergeant at the time I went through training as a "Student Detachment Commander," and had had fifteen or twenty Marines between the ranks of Sergeant and Lance Corporal assigned to my "Detachment." They came from a diverse background to be sure – grunts, box kickers, paper pushers, data dinks and wing wipers to name a few – and before long I was learning the strengths and weaknesses of each.

One of the students in my group always seemed to be sweating, especially during PT, and every time I turned around he was wiping the sweat out of his eyes. I made it a point to stop this practice immediately (since a Marine can't move while standing a ceremonial post), and told the young fellow that if I caught his hands near his face I would amputate them at the wrists! His answer was, "Okay, Gunny... but it stings when it gets in my eyes!" (For the record, he was an Air Wing Admin type...).

Training moved along through a logical sequence over the

next few weeks, with one of the highlights being "law enforcement training" consisting of such things as the PR-24 side-handle baton, pressure point control tactics, and the 12-gauge shotgun – and before long we reached a segment *none* of us had been looking forward to.

The segment I am referring to is training in the use of OC – the dreaded pepper spray. It is made of ground cayenne pepper, is far more potent than riot control agents like tear gas, and can stop a man in his tracks. A formidable weapon to be sure, but there was one catch - in order to be certified to carry the stuff, we had to allow it to be used on *us*. The rationale, according to the school staff, was we needed to know what it felt like in order to be able to use it correctly. The consensus among the class was, "Yeah, but we also carry .357 magnum revolvers – does that mean they have to *shoot* us too?"

When the time came we lined up with our eyes wide open, and one by one the instructors sprayed a stream of OC directly into them. The effect was immediate, and excruciating. To me, it felt as if my eye sockets had been filled with broken glass, and it took the better part of an hour to wash the stuff out using buckets of water provided for that purpose.

Once I was able to see again I started checking on my troops, and naturally the first one I encountered was the "sweater." He was temporarily blind - but I soon discovered that his hearing was working just fine.

"Smith, what the *hell* did I tell you about putting your hands near your eyes!?" I shouted from a few inches away. I'm sure he was thinking, "You've *got* to be kidding me?" – and so after a few seconds I put my arm around his shoulder and said, "Relatively speaking, the sweat's not so bad, is it?"

"No, Gunny!"

SEE PARIS AND LIVE

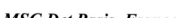

MSG Det Paris, France

W. V. H. White

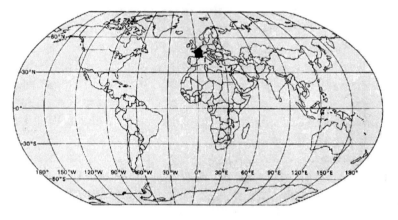

This is a good example of what life is like at a so-called "luxury" post. Unfortunately, a lot of the applicants for MSG duty think <u>all</u> of the posts are like this – and get a rude awakening when faced with the reality of being assigned to Djibouti or some other such garden spot.

There is an old joke in the Corps about the applicant who, being told by a recruiter that he would be going to Parris Island, replied, "Great. I've always wanted to go to France." While it is not at all unusual for today's Marines to visit France while on deployment to the Mediterranean or other parts of Europe, only a select few are assigned to duty in Paris - and nearly all of them are Marine Security Guards serving at the American Embassy.

71

What is duty like in Paris, quite probably the most beautiful and romantic city in the world? It depends on whom you talk to. For some it is a dream come true, while others actually prefer to serve in a comparatively backward, third-world country like many of the poverty-stricken countries of Africa. And both types of individual feel their reasons are valid.

As for Paris, many of the attractions are quite obvious. There are large numbers of historical and cultural sites to be found in and around the city. In addition, the short distances and excellent roads and transportation systems make visiting other areas of France and many other European countries a breeze. Popular day trips are to Normandy, Versailles, Fontainebleu and of course the battlefield at Belleau Wood, a mecca for Marines.

Those embassy Marines who try hard enough can become conversant, if not fluent, in French. Military history buffs are able to study great battlefields from World War II back to antiquity, more than they'll ever have time to see. On the other end of the spectrum, Paris is an art lover's paradise, with the Louvre, the Orsay Museum, the Invalides and some one hundred other museums, two hundred art galleries, and a steady stream of temporary exhibits. Chateaux, palaces, castles and magnificent cathedrals dot the countryside throughout the country.

What functions are performed by the Marine Security Guards while on duty in Paris? Other than a more abundant ration of ceremonial appearances, the duty is essentially the same as in other embassy posts throughout the world. MSG Paris also is a larger detachment than most because it has more posts to man.

The detachment has a table of organization strength of two staff noncommissioned officers and twenty-four watchstanders. The actual strength varies. When this story was being researched, they were short seven Marines, but replacements were on the way.

Operational control of the detachment comes under the Department of State's Regional Security Officer, Andy Colantonio. The Marines stand seven posts, not all of which are manned for twenty-four hours. The posts are located in the embassy itself, and in three other embassy buildings. Access into three of the four buildings is controlled by Marines.

The norm for the duty section is eight hours on and sixteen off. In addition, there are ten days a month off. It is possible to take three or four days and visit such places as Brussels, Belgium; Dublin, Ireland; or London. The guard posts are all interior, with local private guards handling the exteriors and French gendarmes around the perimeter.

The land occupied by the embassy is some of the most coveted in the world. It lies between the luxurious Crillon Hotel and the Presidential Palace. The Place de la Concorde is to the left front of the embassy, and in a couple of minutes you can be strolling to the right along the Champs Élysées toward the Arc de Triomphe. To the left you can walk through the Garden of Tuileries down to the Louvre.

The noncommissioned officer in charge of the detachment is Master Sergeant Richard Doxtader, a combat engineer by military occupational specialty. Born in Michigan, he grew up in Phoenix. He reported to Marine Security Guard School at Quantico, Virginia from the 1st Combat Engineer Battalion at Camp Pendleton. His first post was Tel Aviv, Israel in July 1996, followed by Paris in September 1997.

"It's the best duty in the Corps. I wish I had come out sooner," he said. "It's the only billet in the Marine Corps where a staff NCO is the commander. Staff NCOs who want to go on MSG duty should have some computer skills, administrative knowledge and leadership. Forget the 9-to-5 job, you're always playing catch-up, particularly if it is a larger, more complex detachment," the 'Top' said. "You have a lot of balls in the air, and you have to know which ones not to drop."

MSGs of the Paris Detachment at the American Cemetery near Belleau Wood during the annual memorial ceremony.

For those with children, the schools throughout most or all of the [MSG] program are excellent. They are usually small schools, and they follow a curriculum established by the Department of Defense.

"For the younger Marines, I believe the biggest thing about MSG duty is the travel. They are ambassadors in blue who represent the United States wherever they go. While on the program they should easily be able to get an [associate of arts] degree," Top Doxtader said.

He added, "It is also possible for them to save $10,000 to

$20,000 while on the program. At some posts they can live mainly off of the COLA [cost-of-living allowance] and save the rest. The COLA here is higher because of the very high price on goods and services. For example, the cook at the Marine House, who might be paid $50 a month at some posts, costs $2,400 a month in Paris."

The bottom line, according to MSgt Doxtader, is that MSG duty is a great opportunity for Marines who are mature and motivated. "You come out of school motivated, but there are a lot of distractions, and you have to remain focused," he said.

Top Doxtader and his assistant detachment commander or "A-slash," Staff Sergeant Kenny Hopkins, and their families live in the State Department's Neuilly Compound, about three and a half miles from the embassy. The remainder of the Marines live in another State Department compound across the road from the famous Bois de Boulogne, a huge wooded park on the outskirts of Paris, to the west. It is about five and one half miles to the embassy and can take from fifteen to forty-five minutes to drive, depending on the time of day and traffic.

You might think that the MSGs would come mainly from infantry or military police fields, but they don't. You are likely to find most military occupational skills represented on embassy duty.

Sergeant Ryan Faught of Medina, Ohio, is a HAWK missile operator who came to MSG duty from Marine Corps Air Station, Yuma. At the time of the interview he had about ninety days to the end of his enlistment and planned to enroll at Kent State University in Ohio to study computer illustration. While on embassy duty, his primary assignments were in Riyadh, Saudi Arabia and Lisbon, Portugal. He wrapped up his tour with a temporary additional duty assignment to Paris. He said, "When I go to school, I will use the time management, leadership and organizational skills I learned in my four and a half years in the Corps."

Sergeant Barrett Broyles' military specialty is traffic management. He entered the MSG program from Headquarters and Service Battalion, Marine Corps Base, Okinawa at Camp Kinser. His first assignment was to Kathmandu, Nepal, a post with a detachment commander and five watchstanders. Paris was a huge change.

Sergeant Broyles has the critically important additional duty (which rotates periodically) of noncommissioned officer in charge of the mess. He is responsible for collecting the mess bills each month, which cover the cook's salary and his local taxes, food costs and other incidentals. The mess budget is about $5,000 per month compared to $500 at Sergeant Broyles' last assignment. His duties also include making out the menu following the applicable order from Marine Security Guard Battalion in Quantico. "It's pretty tough in a large detachment, knowing what they like and with people changing continually," he said. Help is provided by two assistants and a mess advisory board, which larger detachments have.

Milk and produce are delivered locally while a supply run is made twice a month to SHAPE (Supreme Headquarters Allied Expeditionary Powers, Europe) near Brussels, for the majority of other food items and other household supplies.

Who does the cooking at Marine House? The chef is Lahadi Rouabah, who is in his third year. Prior to that he cooked in "D" Building, another embassy property. With a chef skilled in preparing many great French dishes, what's on the menu? Mostly American dishes such as chicken nuggets, pizza, etc. "Breakfast is to order, lunch is to order, and the evening meal is from the menu, but is American," he said. "I am very proud to work for the Marines," said Rouabah. "Everyone in France knows they are the finest military service in the world."

The duty is very demanding, according to Sergeant Broyles, with many restrictions. The hardest part is keeping focused. "But for Marines who want to see the world, it's the best," he said. "The travel is unbelievable." Since entering MSG duty he has been to nineteen countries on personal travel or on duty. Since arriving in Paris, in addition to travel in France and Belgium, he has visited Geneva, Switzerland and London, as well as Berlin and Ramstein, Germany.

Paris is busy with a steady stream of visitors including the President, Secretary of State, the Commandant of the Marine Corps and many other dignitaries.

"One thing about the program is that it can give you a dilemma. Really, the experience you gain through working with the diplomatic community gives you an edge going into another governmental field. I love the Marine Corps, but I am looking at my options," Broyles said.

Marine House is one of three apartment buildings in the embassy compound. The other two are for embassy personnel and their families. It is not luxurious by any means, but appears to be very comfortable and functional. There are four floors with the ground floor having a TV room, laundry with washers and dryers, billiards and game room, bar and movie room, and a weapons simulator (9-mm. pistol, shotgun and M16 rifle) for training. The first floor (our second) holds the dining room, galley, gym and two apartments of four bedrooms each. The upper floors are devoted to apartments and a storage area.

Marine House hosts several functions during the year for embassy personnel and their families. Events such as a Fourth of July barbecue with volleyball, children's games, etc., which drew more than two hundred people, have been popular.

Transportation to and from the embassy is no problem. An embassy vehicle and five drivers provide a ride twenty-four hours a day.

Corporal Melisa Laredo was the only woman assigned to the detachment at the time. She is an administrative clerk by MOS, and entered the program from First Force Service Support Group at Camp Pendleton. Paris is her first post. Why did she choose MSG duty? "I always wanted to travel. I have been to Hong Kong, Beijing, Germany and Belgium. It's great duty being able to experience different cultures and interact with people. And, here, I have all my 'big brothers,'" she said.

"You need to be able to adapt well, particularly with no other females. It can be lonely. If you need that, you shouldn't go, but if you can adapt, it's great," she added.

"Here, I'm doing things I've always wanted to do. Off duty, I take dance lessons and acting. For my second post I'd like to go to South America. I would like to learn more of my heritage and improve my Spanish," she said.

The detachment's A-slash or assistant NCOIC is Staff Sergeant Kenny Hopkins, a traffic management operations chief by MOS. Basically, his duties pick up at wherever Top Doxtader's stop. When the Top is away, he picks up his duties also.

"It's demanding, but rewarding also, great duty for a staff NCO. I highly recommend it to all Marines. You have to have a positive attitude and be able to adapt to the environment you're in. The duty presents challenges, but many rewards also."

Staff Sergeant Hopkins and his wife, Larea, a former Marine who had five years in the Corps, live in a three-bedroom apartment that they feel is quite nice. Staff NCOs in Paris can count on lots of visits from family and friends. They make periodic runs to SHAPE for commissary and PX items, and dental and medical needs are also treated there - although using the American Hospital in Paris is an option.

He believes the biggest advantage of MSG duty is being able to see parts of the world other Marines don't get to see. Also, there is much more responsibility than on normal duty assignments.

"Some of the Marines interact with the locals and others with the diplomatic community. But it's all educational and worthwhile," he said.

Travel opportunities are abundant. He and his wife went to Biarritz over the Labor Day weekend. "We have also been to the south of France, England, Ireland, Spain and to Brussels," he said. "If you can stay balanced, you can't match the duty."

There is always an element of danger on embassy duty. This was brought home sharply to Sergeant Shawn Jackson last year. Jackson is an artilleryman who was stationed at the Marine Corps Air Ground Combat Center at Twentynine Palms, in "India" Battery, 3d Bn, 11th Marines, when he reported to MSG School. His first post was Nairobi, Kenya in March 1997, followed by assignment to Paris in May of 1998.

"In August I had just returned from leave, and they told me my prior embassy had been blown up and that I was being sent down there to help out. I was really worried, because I knew everyone there. I was sent first to Dar es Salaam, Tanzania for about three days, and then to Nairobi.

"It was very hectic, a cold, eerie feeling. The diplomatic community really came together, but it wasn't the same. It was no longer the beautiful Kenya it had been before. I would rather serve in a third-world country. The American and diplomatic communities are much closer. You develop really close friends. I lost several friends in the bombing.

I learned a lot in that country. Africa is different in many ways from what you see on TV. It has much to offer if you look for it," he said. "MSG is a great opportunity to advance yourself, and you learn that the world doesn't stop at the borders of the U.S. It's what you make of it.

All of Nairobi was affected. All of those there suffered. All helped to recover. It was a good feeling to see the other embassies fly their flags at half-mast, out of respect," he concluded.

And there you have it - from the beauty and joie de vie of Paris, to the suffering in Nairobi and Dar es Salaam in the aftermath of the bombings. It pretty well sums up embassy duty. There are the glamour, excitement and adventure of travel and assignment to exotic lands - but always lurking in the background is the inherent danger of being out on the front lines of democracy.

That's why U.S. Marines were chosen to guard America's embassies.

ANIMAL HOUSE
Revisited

MSG Detachment Canberra, Australia

One of the best things about MSG duty is the Marine House. The Marine House is also one of the worst things about the program. It's the best because the living standard is (usually) much better than that of a BEQ in the fleet, and I say worst because there is a lot of potential for getting into trouble in such an environment. In the end, it's what the Marines make of it.

One of the unique things about the MSG Program is the way Marines are billeted. Naturally the Detachment Commander is provided with his own house, since only Staff NCOs are allowed to be married while on the program. The rest of the detachment lives in a facility known as the Marine House. For 1/5 Dets (one SNCO and five watchstanders) that

81

usually ended up being nothing more than a big house, but for the larger detachments the Marine House could sometimes resemble a small college dormitory.

One of the big challenges for a Detachment Commander is enforcing the understandably unpopular battalion regulation which prohibits members of the opposite sex (in most cases that means women) from entering the rooms of individual Marines. It is a challenge because it is essentially a battle between regulations on a piece of paper, and the raging hormones of a twenty-something Marine - and Mother Nature can be a pretty formidable adversary.

A typical Marine House - note the Detachment vehicle parked in the foreground.

For the most part that rule wasn't a big problem for my detachment in the Congo, and I am willing to bet the same was true for Marines assigned to many other third world posts. Aside from the occasional airline stewardess in transit there were virtually no women to be found in Brazzaville - at least none who spoke English, bathed regularly, or had not contracted the AIDS virus – so it was a non-issue.

Canberra, Australia was a different case altogether. *All* of the Marines had girlfriends there – and some of them had *more* than one. It was a situation ripe for disaster. Even though I made a concerted effort to make as many unannounced visits to the Marines' quarters as possible, and made it quite clear that I had a zero tolerance policy for transgressors, I knew more had to be done. I simply couldn't be there all of the time. The solution I decided upon was to make my A-Slash (Assistant Detachment Commander) personally responsible for any and all violations, and to his credit there were none (that I knew of).

Despite my stand on the enforcement of battalion regulations, no one can ever say I didn't encourage my Marines to have a good time within the boundaries of the rules. Parties were thrown at the Marine House on a monthly basis, and some of them were quite memorable. They were so good, in fact, that we began to receive regular invitations to attend functions hosted by the Australian military and police, and ended up developing a very cordial relationship with each of those organizations.

One incident that stands out took place early in my tour in Canberra. It was the tail end of one of our monthly parties, and as the crowd thinned out I noticed one of those who remained was an Australian girl who worked in the embassy. I had met her on a couple of occasions and she had impressed me as being quite intelligent and witty, but unfortunately that was not the case on this particular evening. She was, as they say, a bit "green around the gills."

When it came time for her to go home it became apparent she could not even walk, so it became necessary for me to carry her outside to the cab her friend had called. When I placed her in the back seat I thought that was that - but I was wrong.

I felt bad for that poor girl, I really did - but if I had known what she had done I probably wouldn't have been quite so sympathetic. During the course of the party she had been drinking shots poured for her by one of my Marines, who was no doubt doing his level best to get her drunk. He succeeded. When all of that liquor kicked in that nice girl decided she needed to locate a bathroom in order to perform what the Aussies sometimes refer to as the "technicolor yawn." She somehow managed to get there alright, but couldn't find the light switch – so instead she just felt around in the dark for something porcelain with a lid, and that's where she proceeded to deposit the contents of her stomach.

Well, the following morning one of my Marines went downstairs to do a load of laundry, and you can imagine his surprise when he opened the lid of the washing machine to discover the mess left behind by our guest from the night before. When that nice girl was informed she had mistaken the washer for the toilet she was mortified, and promptly sent a bouquet of flowers to the Marine who had made the discovery (and who had cleaned up the mess) along with a written apology. I guess in the end the impression she made wasn't all that bad, since she ended up marrying my A-Slash a couple of years later!

STATE DEPARTMENT
Marines

MSG Dets Baghdad, Iraq and Beijing, China

Major H. G. Duncan

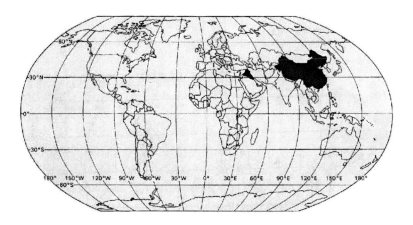

In 1979, for the second time in a year, the American Embassy in Teheran, Iran was surrendered. Up until that time, I can't think of a single Embassy ever surrendering except in times of war when belligerents took control of each other's embassies.

A great percentage of Americans are unaware of the fact most American Embassies have Marine Security Guards assigned to them for that purpose – to provide security. Such duty is considered to be a choice assignment among Marines, particularly the younger ones.

The young Marine conjures up visions of duty in

Stockholm, swimming naked with buxom beauties on nude beaches. He requests embassy duty and goes through the grueling process of qualifying – all the time in a state of tumescence – while thinking of the blonde, blue-eyed Swedes. Then he gets his first assignment – to Haiti, Liberia or Afghanistan. It comes as quite a shock that there are embassies other than those in London, Paris and Rome!

In my younger days I got an embassy assignment... almost. In 1952 I had completed the Russian language course at the Naval Intelligence School in Washington D.C., and since there was a need for more Marines in Moscow I applied, was accepted, and reported to Henderson Hall for processing and a long wait.

Our relationship with the Soviet Union was strained at that time. Any little thing the Soviets could do to irritate we Americans was relentlessly pursued. The routine relief and assignment of Marines to our embassy in Moscow was one of those irritants. In fact, I met a Marine at Henderson Hall who had been waiting for seven months for clearance – so I started looking for a way out of my assignment, Fortunately I found one, and was never honored to serve on embassy duty.

A good friend of mine was, however. Phil Smith, who later became a Sergeant Major, served in his younger days as a member of the MSG Detachment in Baghdad. Those middle-easterners were not very friendly toward us, but tolerated our embassies. One day Smitty and another Marine were on liberty in Baghdad when a huge oil refinery exploded and burned. The two were arrested on the fringes of the conflagration, and accused of a capitalist plot to destroy the country. They stayed in jail for quite a while before the diplomats could negotiate their release and get them returned to the United States. I later asked Smitty if they had anything to do with the fire, and he coyly denied it.

Raising the flag over the new American Embassy in Baghdad, Iraq.

Years later aboard Camp Pendleton a Gunnery Sergeant came to me for counseling on a fitness report he had received upon detachment from his last duty station, an embassy where he had served as the Commander of the MSG Detachment. The reporting senior had been the embassy RSO, who happened to be a woman. Since it was a derogatory report, it had been sent to him for review and rebuttal. I had never seen a report quite like it. He was marked "Unsatisfactory" all the way down the line. No one is *totally* unsat! I finished reading the report and said, "Sounds to me like you should have kept on screwing her."

He told me that he had indeed romanced the security officer to the extreme, but when he found and married the woman of his dreams the RSO was forced to step aside. The Gunny also told me all of his previous reports from her were on the other end of the spectrum – "Outstanding" across the

87

board. It just goes to show that women in positions of authority are no different than men where such things are concerned.

In 1966 in Alameda, California I influenced a really fine young Marine to reenlist. He wanted embassy duty, got it as his reenlistment option, and was assigned to Saigon, South Vietnam where he became one of the few Marines – in modern times, at least – to draw combat pay while wearing dress blues.

In the days of Nixon we reopened talks with Communist China. President Nixon went to Beijing (then Peking), and obtained authorization from the Chinese to open an American mission there. A small detachment or Marines was assigned to guard it – all of whom with recent combat experience in Vietnam. Right away the Chinese began complaining about the Marines wearing their combat decoration in town, because it was "offensive to the people," and as a result an order was issued that uniforms would not be worn outside the legation. The Detachment responded by falling out for PT every morning wearing their scarlet and gold USMC athletic gear, and their runs through the People's Park generated other complaints – but the real clincher was the "Red Ass Saloon."

The Marines, restricted from wearing their uniforms in public and limited as to where they could conduct PT and in what clothing, decided to open their own bar. It was located on the second deck of the detachment quarters in a small, sparsely decorated room that contained the bare necessities – booze, and a place to sit and talk. The Chinese were unaware of the establishment, and would have remained ignorant had the Marines not become rowdy one particular night. In the end it was just another excuse for political gamesmanship.

Yes, it was probably a good thing I did not end up on

embassy duty. I was not to sort to conduct myself with restraint, and my idea of a good MSG was best captured in a televised news clip from one of our embassies in Central America around the time of the Iranian hostage mess. It showed two young Marines from the MSG detachment on the balcony of the embassy, dressed in jungle utilities and armed with shotguns. One Marine stood out in particular. He was about nineteen or twenty, and had no cover on his head. His blond hair was mussed, he had a cigarette dangling from his lips, and held a shotgun at his hip. The determined look on his face seemed to say, "Come on, you bastards! Just try and get by me!" He is the kind of Marine I like to remember, and the sort we should assign to such duties. Even so, he probably got his butt chewed for not wearing a helmet and smoking within range of s camera – and for thinking obscene thoughts!

THE MESSAGE

MSG Detachment Brazzaville, Congo

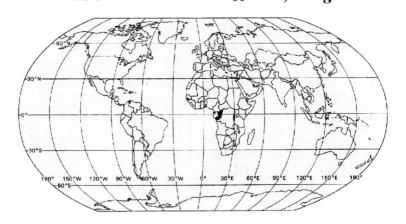

There are a lot of different personalities in an MSG Detachment, as there are in any group, but they are often more noticeable because of the constant contact the members have with each other (remember, they live in the same house in addition to working together) and the pressure-packed situations they must sometimes face. Because of that, a Detachment Commander must be more than a tactician, a supervisor, and an administrator – he must be a psychologist as well – and sometimes that requires him to use a bit of humor.

One of the characters I had working for me when I commanded an MSG detachment in the Congo was a Marine by the name of Ruben Castillo. Sergeant Castillo was a tough Chicano from Chicago, and one of the most competent watchstanders in the detachment. He was also a bit of a wise-guy.

One of Castillo's pet peeves was the habit Congolese "guardians" had of falling asleep on the job. The guardians were uniformed security guards hired by the Embassy to provide security for our homes, and there was one assigned to every American residence twenty-four hours a day. One such fellow was posted at the Marine House, as the detachment "dormitory" was known, and had been seen snoozing on numerous occasions. He always claimed that he was simply resting his eyes for a few moments, so my sergeant decided to prove him a liar.

Late one evening Castillo covered his face with green camouflage paint and gained entrance to the Marine compound by slipping over the eight foot wall surrounding the residence. He crept around the side of the building, and spotted the guard in his customary location – fast asleep in a chair on the front porch. To prove his point Castillo pulled a can from his pocket and covered the sleeping guard in shaving cream, after which he woke him up. When confronted with circumstances that could lead to his getting fired (jobs at the Embassy paid relatively well and were coveted by the locals) the man swore he was awake the entire time and had *allowed* himself to "slimed." Believe it or not that was his story, and he was sticking to it!

One of Castillo's personality traits was his stubbornness. He was worse than a mule. For example, one day I decided it was time for everyone to change their radio call signs in order to enhance communications security, and being the nice guy that I am allowed each Marine to choose his own. I even gave them a week to submit it to me for inclusion in the Embassy list - but Ruben had some macho handle that he really liked, and I guess he figured if he didn't give me a new one he could just keep it. Wrong answer. I *assigned* one to him, and designated his call sign to be "Plant Man" because it reminded me of how he had looked the night of the shaving cream raid. He initially refused to answer to it, but he and I got that problem straightened out in a *hurry*.

Prior to coming to the Congo Castillo had served at the Embassy in Paris, and while there had met and become engaged to a lovely English girl. At one point she came down to visit him for a week or so, and we all decided she was *much* too good looking for him - and she was nice to boot.

As Sergeant Castillo neared the end of his tour in Brazzaville, and on the MSG program for that matter, he and his fiancé began making their wedding plans. When his tour in Brazzaville ended he was to fly straight to England for a big wedding, and from there they planned to go to Aruba for their honeymoon before he reported in for duty at Camp Lejeune.

When I heard the details of the wedding and honeymoon, and found out that the invitations had been mailed and everything paid for, I decided it was a perfect opportunity to give the sergeant a taste of his own medicine. Most of our official directives came via naval messages, which were processed upstairs in the communications center and placed in our box each morning. These messages have a distinctive format, and anything contained in one is treated as if it came from the burning bush. So I drafted a bogus message of my own, and had the State Department communicators run it through the system so it appeared to be legitimate. I then placed it in our in-box, and waited for Castillo to take the bait when he picked up the morning traffic.

The message simply stated that any MSG who had less than a year remaining on his enlistment, and who had either not submitted a reenlistment request or had submitted one that had not yet been approved (Castillo fell into the second category) would be required to remain at their current post until approval was received or they reached their EAS.

When I arrived at the Embassy the next morning I knew the ruse had been successful because Castillo was in what can only be described as a highly agitated state. I had barely walked through the door when he shoved a copy of the message in my direction, followed by a string of expletives. I could barely keep a straight face.

The Stone Age telephone service we had in the Congo was, for once, a good thing. Castillo had tried calling back to MSG Battalion, Headquarters Marine Corps, and even his Congressman. Fortunately he got through to no one, although he vowed to keep trying until he did. I sat in my office trying not to laugh while he paced back and forth on Post One, ranting and raving. It was fun to watch, but I didn't dare leave him alone in such a state.

Finally I walked into Post One and asked Castillo for the message. When he handed it to me I ripped it into small pieces and said "There, that takes care of *that* problem!" Castillo stared at me in disbelief, since he figured he would need the message as a reference. "You can't do that Gunny," he exclaimed. "I need that!"

As I headed out the door I tossed the message into the trash can. "Sure I can," I said. "I wrote it!"

PERSONA NON GRATA

MSG Detachment Canberra, Australia

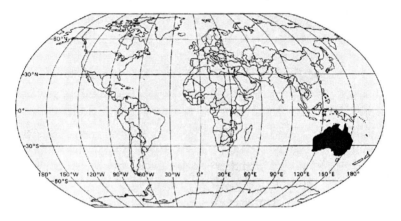

Sometimes it's easy to lose sight of the fact that most MSGs are young and relatively inexperienced in many areas, because of the tremendous amount of responsibility placed upon their shoulders and their "can-do" attitude – but every so often one of them would say or do something to remind me!

One of the jokes the other services like to tell at our expense is that 'MARINE' means "Muscles Are Required, Intelligence Not Essential." What those detractors often fail to realize is Marines are trained to obey rules and regulations, and what could appear "stupid" to an outside observer might simply be a Marine following the rules – as he understands them to be.

Marines who serve at our Embassies abroad are known as Marine Security Guards, or MSGs, and those who go on the program must go through a rigorous screening process - and with good reason. At most posts around the world, a single Marine is often responsible for the security of the Embassy or Consulate at any given time, and above-average intelligence is required in order to make critical decisions in the face of a variety of unexpected emergencies. Attacks have been launched, bombs have been detonated, and fires have broken out in the blink of an eye – and in each case the State Department employees in the building look to the Marine on Post One for the appropriate response.

Parliament House, seat of the Australian government.

Because of the exceptionally high caliber of Marines being sent out to MSG posts, I sometimes forget that while they are certainly squared-away and competent watchstanders, these guys are also young and still a bit naive in some matters. I would always be amazed when a young hard-charger who was capable of programming a computer with his eyes closed would turn around and say or do something absolutely outrageous.

A good example occurred while I was commanding the MSG Detachment in Canberra, Australia. Unlike many of the third-world or "hardship" posts we had to serve at, Canberra is a modern and cosmopolitan city - and we often took advantage of that by dining in the many eating establishments located in the vicinity of the Marine House. In fact, we dined out so often that in a matter of months my Marines had eaten in each of the restaurants several times. So it came as quite a surprise to me when I discovered one of my guys had never eaten in the Chinese restaurant located right next door to our favorite watering hole. I knew for a fact he liked a variety of foods, so I asked him why he hadn't gone there. His response was, "I thought you knew, Gunny. We're not *allowed* to eat in there."

I was dumbfounded by that, and ready to march down to the restaurant and "read the riot act" to the restaurant owner. But before doing so, I wanted more information.

"Who *told* you that you can't eat there?"

"Well, nobody. But there's a sign on the door."

"A sign? What does it say?"

"No MSG!"

Needless to say, I proceeded to explained to the young lad about Monosodium Glutamate – once I finished laughing.

FLY THE FRIENDLY SKIES

MSG Detachment Brazzaville, Congo

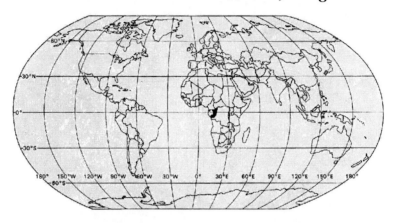

I have often said that I spent my entire Marine Corps career waiting for something to end - be it a night's duty, a week in the field, a month aboard ship, or a year in Okinawa – and my tour of duty in the Congo was no exception. There were some great moments to be sure, but two years (I was extended twice) is a <u>long</u> time to be in a place like that!

Whenever someone finds out that I made a couple of hundred parachute jumps during my time in the Marine Corps, the typical reaction is "Oh my, that sounds awfully dangerous. You'd never catch *me* doing that!" It's all I can do to keep from laughing. Granted, throwing oneself from a perfectly good airplane is not a natural act, and yes, it can be quite exhilarating. And of course there *is* the occasional unfortunate accident – but all-in-all the idea of jumping scares me far less than many other things. Take driving for instance. Have you *seen* some of the people out on the roads these days? It's downright frightening! And almost as frightening is the prospect of taking off and landing in some of those "perfectly good" airplanes mentioned earlier. Think about it for a second. You are entrusting your life to a pilot you have never even *met!* And odds are the aircraft's maintenance has been performed by a high school dropout whose uncle got him into the union in exchange for a kickback to the shop steward. Not too encouraging, is it?

Maya Maya International Airport, the only port of entry into Brazzaville, Republic of the Congo.

In all fairness though, the safety record of U.S. flag airlines is pretty good – although not perfect. I am reminded of the autistic character played by Dustin Hoffman in the film *Rainman*, who will only fly on Qantas Airlines because they have never had an accident resulting in a fatality. Unfortunately, the Aussies don't fly to all destinations, and for that matter neither do U.S. airlines. It is therefore occasionally necessary to fly on foreign flag carriers, which can sometimes be quite an adventure.

My first experience with third world airlines came when I was traveling from Camp Lejeune to my first Embassy posting in the Republic of the Congo. I had been warned by those 'in the know' to avoid flying Air Afriqué at all costs, and I made it a point to specifically request a ticket on one of the two other airlines serving Brazzaville, namely SwissAir and Sabena. Naturally, my request was ignored. When I changed planes at Charles De Gaulle Airport for the flight from Paris to Southwest Africa I was greeted by the ugliest puke green jet I had ever seen, and knew immediately it was going to be a long, long flight.

The thing about that flight which sticks out most clearly in my mind was the food they served. I remember we had a choice between fish, and something else which I could not identify. The fish smelled like it had been rotting in the sun for a week, and the other choice looked like a piece of roadkill. I declined both.

As if the menu wasn't bad enough we had not one, but two, stops enroute - in places not known for their amenities. The first, N'djamena, Chad, was a charming example of North African hospitality. From the air the place looked like nothing more than a collection of poorly constructed mud huts, and the desert stretched in all directions for as far as I could see. When we landed our aircraft was immediately surrounded by a horde of AK-47 toting troops, and for a time it looked as if we were going to be boarded. After a tense hour or so in the middle of the runway we finally deplaned a few people, took on a couple of passengers, and departed.

The next stop was Bangui, capitol of the Central African Republic – a world class sh*thole to be sure, but a lush tropical paradise in comparison to N'Djamena. From there it was on to Brazzaville.

Our flight finally arrived in the Congo late in the evening, and although I was initially glad to finally be in Brazzaville and at journey's end, that soon changed. They packed us onto a bus for the trip from the plane to the terminal (there were no jetways), and I got my first real taste of "olfactory overload." That is what happens when one's sense of smell is overwhelmed by a particularly repugnant odor – in this case a busload of people who think good personal hygiene means bathing once a month, whether they need it or not. I fought to get a window open, but to no avail.

Upon arriving at the terminal (by this time it was the middle of the night) I discovered it was the custom in that part of the world to pay the AK-47 toting customs official a bribe in order to get one's passport stamped. At least that's what I found out once the "expediter" arrived from the Embassy and translated for me. It was a harbinger of things to come for the next two years.

A few months later I was traveling on an Air Cameroon flight from Nairobi, Kenya back to Brazzaville, with a stopover in Bujumbura, Burundi. I expected it to be pretty much like any other flight – right up until the moment one of the other passengers tied the leash of his goat to the armrest of my seat. That's right... I said GOAT. I asked what the heck was going on, and was politely informed that passengers were allowed to travel with a domesticated animal in lieu of a piece of carry-on baggage. "How cosmopolitan," I thought to myself.

When it came time to eat, the flight attendant came around with a tray of sandwiches from which we could choose. To say they were unappetizing would be a gross understatement. The sandwiches they served were reminiscent of a scene from "The Odd Couple," where Oscar's friends are rummaging through his refrigerator during their weekly poker game and discover there is a choice between brown sandwiches and green sandwiches. When asked to identify the green sandwiches, Oscar tells them they are either "very new cheese, or very old meat." Naturally, they select the brown ones – and I did the same. I ended up feeding it to the goat.

Then came our stop in Burundi. We picked up a few passengers, which is not unusual. What *was* unusual was the cargo we took on. It was an SUV. When the plane began to rock violently back and forth I peeked through a crack in the bulkhead, and saw that they had actually opened the side of the airplane to put it in. The allotted space was just large enough to hold the vehicle, and several laborers in native dress (i.e. nothing but loincloths) were trying to slide it in sideways by picking up first one end, and then the other, over and over. The weight distribution was enough to make the loadmaster on a C-130 cringe. I wasn't too confident about the airworthiness of the plane at this point, but just as I was contemplating getting off and making my way back to the Congo through the jungle (and the warring Hutu and Tutsi tribes) we buttoned up and took off. For the remainder of the trip, I drank heavily.

When I finally completed my tour in the Congo I was in a position to make my *own* reservations, and my trip out was a darn sight better than the trip in on Air Afriqué. I was given a seat in first class by my friend Francois, the local representative for the Belgian airline, Sabena. To this day it remains the only time I have flown in a class other than cattle car economy, and I was determined to enjoy it. As I sat back in my plush, oversized seat and sipped a celebratory glass of champagne it suddenly occurred to me that even though there was no goat tied to my seat and the brown and green sandwiches of Air Cameroon had been replaced with fine Continental cuisine, there was still the possibility that we could end up in a twisted mass of wreckage in some remote jungle.

I found myself missing my parachute once again.

OUR MARINES

MSG Det Madrid, Spain

Colonel Edward F. Danowitz USMC (Ret)

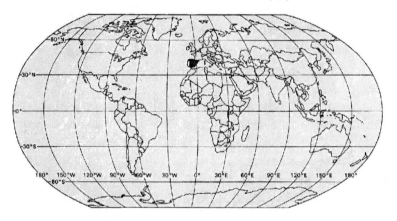

They say that one of the seven deadly sins is Pride, and if that's the case we Marines are all doomed to burn in hell – and that goes <u>double</u> for MSGs. Unlike some Detachment Commanders, I was fortunate when it came to the Ambassadors I served under. Each of them was proud of his detachment and the Corps as a whole (one had even been a Marine himself), and all treated us in a respectful manner at all times.

Marine Security Guards always have had a profound presence at the embassies to which they have been assigned. This was the case when I served as the Assistant Naval Attaché at the American Embassy in Madrid, Spain during the late 1950s.

Our Ambassador, John Davis Lodge, had great pride in the Marines posted at the embassy. On the occasion of a visiting dignitary to the embassy, Lodge would form the Marines to be introduced as important members of his staff. I was also invited to meet with the Marines, in dress blue uniform, and to walk with the guest of honor as each Marine was presented.

On one such occasion I remember well the visit of the Chief of Naval Operations, Admiral Arleigh Burke, to the embassy prior to a call on the CNO of the Spanish Navy, Admiral Don Felipe Abarzuzza, to discuss modernization of the Spanish fleet.

Also present was a Congressional delegation headed by Senators Henry Styles Bridges, Milton Young and Estes Kefauver. They had come to Spain to view the new Air Force bases under construction at Torrejon, Zaragozza and Sevilla, as well as the Naval base at Rota.

Once the Marines had formed, Ambassador Lodge turned toward his guests and to the Admiral said, "Arleigh, I'd like you to meet my Marines."

In immediate response, Admiral Burke replied, "John, you're wrong. They're *my* Marines. They belong to the Department of the Navy."

Before the argument could proceed further, Senator Bridges, the wise lawmaker from New Hampshire, proclaimed in a loud voice, "Gentlemen, you are both wrong." After a pause he added, "They're *our* Marines. They belong to the American people!"

The chests of the Marine Security Guards braced as never before, and deservedly so, because each realized the high esteem in which the other services and the American nation hold our Corps.

MSGs IN ACTION

BOXER REBELLION

American Legation Peking, China

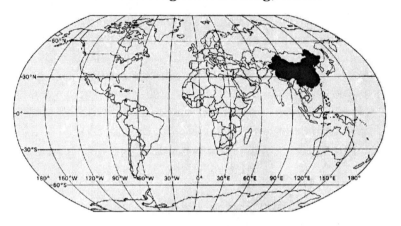

Before the official inception of the MSG Program in the 1940s there were many instances where Marines came to the rescue of American diplomatic posts around the world. One of the most famous took place in China at the turn of the century.

One of the Marine Corps' most famous engagements - the defense of the American Legation at Peking (now Beijing), China, is also one of the most celebrated acts of valor in the history of the Corps. Celebrated in the 1962 Charleston Heston blockbuster movie *55 Days at Peking*, Marine exploits at Peking and Tientsin were highlighted by the heroism of legendary Marines such as Lieutenant Smedley Butler and Private Daniel Joseph Daly.

Butler fought at Tientsin, where he was brevetted for heroism while Daly was among Marines of the American

107

legation guard and saw action when an anti-foreign secret society, the Fists of Righteous Harmony, also known as the Boxers, attacked Legation Quarter, the home of foreign diplomats and their families.

More than twenty thousand strong, the Boxers marched through the gates of Peking unopposed by Chinese Imperial forces and laid siege to Legation Quarter. Daly and a force of forty-eight Marines and three sailors under the command of Captain John T. Myers arrived in the capital on May 31, 1900 shortly before the city was surrounded by the Boxers.

Another detachment of Marines and sailors under Captain Newt Hall, USMC, was assigned to defend the Methodist mission located at some distance from the Legation.

1900: Members if the American Legation Guard, the forerunners of MSGs, on the Tartan wall in Peking (Beijing) China. A handful of such Marines successfully defended the Legation Quarter against tens of thousands of Chinese Boxers in one of the most famous actions in Marine Corps history.

In fighting to defend the Legation wall, Marines fought alongside German troops to repulse Chinese hand-to-hand attacks while withstanding artillery fire. At the insistence of Captain Myers the Americans and Germans deserted the wall and moved back into the legation, and by doing so probably avoiding defeat and a massacre of the inhabitants of the legations. The Boxers next began to rake the defenders with concentrated fire.

The combined national forces held the Legation Quarter under Myers' leadership. Eventually Captain Hall, unable to hold his position at the Methodist church, organized his band of Marines into a surprise attack on the tower on the night of July 2. Hall took the Chinese tower in a demoralizing defeat for the Boxer forces.

Private Daly's Medal of Honor citation states that he distinguished himself for meritorious conduct for acts of heroism on August 14, 1900. The most famous of Daly's now-legendary exploits throughout the July-August defense of the Legation Quarter was during the afternoon of July 13 when German soldiers had been driven back from their position on the east end of the wall. Daly volunteered to take up a position there and provide cover fire while repairs were made to the fortification.

Daly replied "I'm your man" to Captain Hall's request for a volunteer. Daly held his position, alone, throughout the course of the night, withstanding repeated Boxer assaults. Relieved at dawn, Daly was found to have accounted for more than two hundred Boxer dead, subsequently allowing the Marines to reclaim a position the German forces had lost.

Although Boxer assaults continued at other portions of the legation wall, no further attempts were made at the position held by the Germans and Americans along the West wall.

While the Marines successfully defended Peking, a force

of two thousand - including a detachment of one hundred and twelve American sailors and Marines under Navy Captain B.H. McCalla - was assembled from amongst ships of the foreign fleets off the coast and began the long march to relieve the beleaguered international settlement.

McCalla was wounded three times in fighting in Tientsin and was later reinforced by 142 Marines led by Major W.T. Waller, USMC, who led Marines in heavy fighting for Tientsin and later relieved the Legation Guard in Peking.

This action may not have been carried out by Marine Security Guards as we know them today, but the outcome was the same – the Diplomatic Corps was secure, and the situation well in hand!

WITHOUT SO MUCH
As a Bloody Nose

MSG Det Seoul, South Korea

WO George V. Lampman, USMC (Ret)

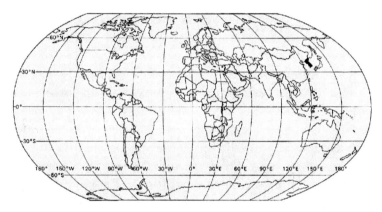

Author's note: This story is a factual and chronological composite of the firsthand experiences of ten of the Marine Security Guards who were there during the evacuation of the American Embassy in Seoul - the first such account ever published. After fifty years, recollections do sometimes blur - but to the greatest extent possible, this story is what really happened. All nineteen members of the Marine Security Section made it safely out of Korea, and Edwards remained with the Ambassador. Seventeen of the Marines retired from the Corps in the late 1960s - one a lieutenant colonel, one a lieutenant, one a warrant officer and the rest senior NCOs. At least seven have passed on, maybe more. It's not certain, since a few have been impossible to locate in recent years.

111

American embassies around the world throughout the years have had U.S. Marines on duty for security and physical protection. The Marines protect the Ambassador and other State Department officials and employees of the embassy, as well as provide physical security of the embassy buildings as required. During the Second World War the requirement for Marines in combat and the closure of many embassies significantly curtailed the program, but subsequent to the end of the war and due to the pressures of the Cold War a formal program was established. In June of 1948, an ALMAR (Marine Corps message) was distributed seeking volunteers.

The requirements were strict: NCO rank with a clean record, unmarried and agreeing to remain so for three years, third-generation U.S. citizen, and eligible for a Top Secret security clearance. Those who were selected received orders to Headquarters Marine Corps at Henderson Hall for duty on "Special Assignment" Overseas with the State Department.

The twenty-man detail selected for the Republic of Korea was an experienced cross section of the Marine Corps in 1948. The senior noncommissioned officer in charge, a master sergeant, plus the two gunnery sergeants and two platoon sergeants were of the pre-Second World War Corps. They all had seen extensive service and action in the South Pacific. The same went for most of the buck sergeants who, by this time, rated two hashmarks. The remainder of the detachment, all junior NCOs, were on their second enlistments. This detail reported to HQMC on 10 November, 1948 and was billeted in "F" Barracks, Henderson Hall.

Training included a brief history of Korea from the ancient to modern era. There were many lectures on security and how to handle what were called "situations," referring to the limits of what could be done in protecting charges and carrying out duties. Training also included weapons familiarization firing and shopping for civilian clothes in downtown Washington. Marine Corps uniforms, ID cards and anything military were left behind. This training was accomplished without incident by 20 December.

On 27 December, Military Air Transport Service flights took the Marines by stages to Kimpo Airfield near Seoul, arriving on 9 January, 1949. The next day at mission headquarters a familiarization tour was given by the then-embassy security officer, Richard Jones. The security officer had no preconceived notions of what the Marines were capable of doing. They had to derive their responsibilities by trial and error to see what worked and what did not. Within two weeks, they had set up a well-oiled system of posts and watches in the mission headquarters.

The diplomatic mission, later to become the American Embassy, was in the eight-floor Bando (Peninsula) Hotel. Built during the Japanese occupation, it was semi-fortified with steel-shuttered strong points guarding the first two floors, and partially surrounded by a dry moat.

Two Marines, Technical Sergeant Jack Edwards and Sergeant Lloyd Henderson, were assigned as bodyguards for the Ambassador and were quartered in the official residence compound in Chung Dung. There was a twenty-four-hour, three-man watch at the mission headquarters. In addition, five Marines were assigned to the assistant security officer to check outside areas during the day.

The Marine control system monitored admittance to the building and a further check of the personnel allowed entrance to the fifth-floor embassy offices and code and file rooms. The Marines conducted security "shakedowns" of office spaces to be sure there were no sensitive materials exposed before the Korean cleaning crews started to work.

The building was under total security control with well-established procedures and requirements.

Military rank had no bearing on the duty assignments and watch schedules. With the exception of the NCOIC and the two Marines assigned permanently to the Ambassador, everybody was equally "in the barrel" for any assignment. This arrangement worked well. There were very few incidents or any major problems. The obligation to protect the staff and property of the embassy was always first in the minds of the members of the detail. Duties became routine as time passed, and nothing really happened until 25 June, 1950.

On that morning there were three Marines standing the 2400-0800 watch. Sergeants George V. Lampman and Augustus E. "Gus" Siefken were two, but after fifty years none of the participants can remember the third – but he was probably Sergeant Duane E. "Dewey" Lowe.

The main offices of the embassy were on the fifth floor of the Bando Hotel. Access to the fifth floor was restricted after working hours, save for one stairwell entrance controlled by the "Blueboy" post - which was the home base and call sign for the limited radio net operated by the Marine security section. A second post screening visitors was in the first-floor lobby. The third post was "Rover," a radio jeep that continually patrolled and visited all the embassy-occupied hotels, warehouses, open storage areas and outlying residences during hours of darkness. A Korean police lieutenant and an embassy interpreter always accompanied Rover. During normal working hours, this third Marine would control entrance to the embassy main lobby. Also on the communications net were the embassy fire chief and security officers. Outside normal working hours the embassy security telephone rang at the fifth-floor Blueboy post. At

0530 Lampman started receiving inquiries from people asking, "What's going on?" He responded there was nothing he knew of, and logged the calls. Sometime after 0600, when Lampman rotated to the lobby post, Jack James, a United Press correspondent, came into the lobby asking the same question. Soon he was joined by Sarah Park, another reporter. The answer was always the same: nothing.

Unbeknownst to the Marines the Korean War had in fact started at 0400 that morning. At 0800 the watch was relieved by Sergeants Paul Dupras and Glenn "Tiny" Green along with Corporal William "Bill" Lyons. Jack James reappeared at 0845 in great excitement, saying, "The North Koreans have crossed over the parallel in force."

Dupras responded at first, "So what? That is a common occurrence."

James said, "Yeah, but this time they've got tanks."

It became apparent that this was the real thing. About 0830 Ambassador John J. Muccio appeared at the Blueboy post and directed Dupras to immediately locate Robert Heavey, the recently arrived embassy security officer. Heavey arrived and, after a short consultation with the Ambassador, told Dupras to recall all the Marines.

The Ambassador's personal security team of Edwards and Henderson, both heavily armed, established a strongpoint outside the Ambassador's office. Green, joined by Corporal John L. "Spanky" Sullivan, who had just arrived, drew M1 rifles from the armory and took up post on the roof. What this precaution was intended to defend against was never clear. As other Marines arrived, they were directed to critical areas in the embassy.

Upon being relieved from the night watch, Siefken and Lampman returned to their new quarters in the Capitol Apartments, just a few blocks behind the Korean capitol building. No sooner had they hit the sack than they were awakened by the room clerk with a "phone call in the office." Dupras was on the line and said simply and forcefully, "Get back to the embassy immediately with any of the guys you can find. And bring any weapons you have with you." A jeep was already on the way. Around 0930, Siefken and Lampman were on the way back to the embassy. Just as the jeep cleared the capitol building and turned right onto Qua Ha Moon, they noticed a formation of what looked like P-51 Mustangs flying very low over the city. They assumed it was the first class of Korean Air Force pilots due back from training in Texas, putting on a show for the local folks. At least they thought so, until two planes peeled off and made a strafing run at the jeep from the 12 o'clock position. To distinguish them from Army vehicles, embassy vehicles were painted bright orange - the perfect target. Needless to say, they screeched that jeep to the side of the street and bailed out behind some stonework. After the two Soviet-built YAKs made a few more passes, the Marines and the Korean driver got back into the jeep and made it to the embassy without further incident.

At the fifth-floor control point, Master Sergeant John F. Runck was mustering the Marine security section. By this time most of the section had reported aboard. Those not on watch were told to go to the embassy main dining room, eat something and stand by while Runck and Embassy Security Officer Heavey made assignments from a pre-prepared order. Meanwhile, the arsenal had suddenly blossomed from the normally carried .38s to a vast array of Japanese carbines, a Thompson submachine gun, a few M1s, shotguns, etc. To this day many of the men have no idea where all this armament came from so quickly. Scuttlebutt was rampant. "This is no drill. The 'Goonies' [slang for the North

Koreans] are on the way in strength." The only question was how soon would they be at the city gates.

One Marine was given the assignment to drive to the main railroad station to meet embassy staffers arriving on the night train from Pusan. He reported a very slow trip - the streets were crowded and the YAKs were strafing indiscriminately, causing panic and civilian casualties. The main streets around the station were jammed with the troops of the Korean Army Second Division, who were being rushed north from their training area near Taejon.

Several other Marines were sent to alert embassy families in outlying, isolated residences that they needed to be ready to move out quickly should an evacuation be initiated. The decision to evacuate dependents was made around midnight on Sunday. Ambassador Muccio's deputy, Everett Drumwright, reported that by that time it was clear the North Korean forces headed for Seoul through the Uijongbu Corridor could not be stopped.

Early Monday morning, Lowe and Lampman were given the task of destroying all of the embassy vehicles in the motor pool which were not running and were on "deadline." With M1s and cases of ammunition, they went to the motor pool. They lifted each hood, aimed exactly into the block just behind the flywheel and put in a couple of rounds. The damage the M1s caused would prevent the North Koreans from cannibalizing any of the vehicles to make others run. Capturing the embassy motor pool would not add to North Korean mobility. Since there were jeeps, three-quarter-ton trucks, sedans, cargo trucks and extra engines in the shops, it took several hours to "execute" the task.

By the afternoon of the 25th the embassy staff, secretaries and code-room personnel had reported to work and were screening classified material to be burned. Although relatively calm on the inside, the building itself was exposed to danger. Embassy staff were told to remain inside and stay off the roof because of North Korean strafing. The Ambassador himself moved to Drumwright's office, as his own corner office was too exposed to random bullets.

As burning got underway, it became clear what an enormous job this would be. The Army attaché alone seemed to have tons of the material, mostly confidential technical manuals. It all had to be destroyed by the Marines, at first using the furnaces in the embassy basement on a day that was already a boiling ninety degrees.

The Marines manned dollies and collected the classified materials which the embassy staff had stacked for them outside office doors. They trucked the material down to the basement, and fed the furnace all day Sunday and into the night. With short, infrequent breaks for food and rest - and sometimes serving as convoy escorts to Inchon and Kimpo Airfield - they hauled and burned and drove from Sunday morning straight through to Tuesday afternoon. They got the job done.

After a short break, the burn detail continued outside the embassy building. Dupras was involved in setting up a homemade cage, made out of cyclone fencing and steel posts, in which to burn documents. This setup, in the parking lot of the embassy, caused a problem when the large, constant fires brought most of the Seoul City Fire Department to the scene. The Marines argued to let the fires burn, and almost lost the argument until an interpreter intervened. From time to time, when they were not on other assignments, the rest of the Marines helped with the burn detail.

Preparations for the evacuation were being handled by other members of the Marine detachment. Krouse supervised a dozen Korean watch supervisors and about fifty watchmen. Arriving at the embassy on the morning of 26 June, Krouse learned of the Ambassador's decision to evacuate all nonessential American citizens, with the Marines to be the last to leave. As the Americans left, Krouse's responsibility was to cover the areas for which the embassy was responsible and ensure the watch force remained in place.

That morning there were South Korean soldiers running through the streets of Seoul in panic alongside the civilians. North Korean planes were strafing the city and Kimpo Airfield. Civilians and demoralized Korean military were heading for the Han River Bridge, crowding across in disarray and struggling southward to escape the city. Maintaining order during the evacuation was not an easy assignment.

From the morning of the 25th through the end on 27 June Krouse was constantly on the move, checking the various watch posts and trying to keep embassy facilities from being pilfered. The various sites were becoming less and less secure because the watch force was leaving its positions. On the 26th alone, half the supervisors left. These men were going home to take care of their families in the face of the invasion. By midmorning on the 27th Krouse was forced to dismiss the remaining watch force and return to the embassy proper where he joined the Marines there already burning classified documents. From the time of the invasion until he and the other Marines were finally evacuated by air, Krouse never left his assigned duties. He, along with most of the Marine Security Guards, left Korea with nothing but the clothes on their backs.

As the evacuation progressed during Monday, 26 June, most Marines were relieved in rotation to return to quarters, pack a few belongings, and return to the embassy. Dupras,

Green and Lyons had a chance to leave the embassy at 0900 for the first time in more than 25 hours. All American dependents (women and children of embassy staff) plus nonessential employees and many other foreign nationals were assembling with their baggage at the embassy. A convoy of buses took these people to Inchon to evacuate. Several people did not want to leave, and had to be forcibly put on the buses. Marines escorted and controlled the bus convoy to Inchon.

The large group of 682 women and children was taken to the port at Inchon and put on the only ship available, a Norwegian fertilizer hauler named *SS Reinholt*. It would not be a pleasant voyage to Japan. The convoy returned Monday evening, and the Marines carried out other assignments.

Late Monday afternoon several Marines escorted another bus convoy of single American Embassy employees to Kimpo aerodrome where the Far East Air Force (FEAF) planned to run in several C-54 transports to begin the air evacuation. There were no ships, other than the Norwegian one, close enough to Inchon to get more people out by sea before the North Koreans were expected to reach Seoul. The remainder of the evacuation would have to be by air or overland. While at Kimpo, this group saw two FEAF F-82s, (twin-boom Mustangs) knock down two YAKs which were attempting to get at the C-54s. The FEAF was able, all during the evacuation, to protect the air bridge taking people out of Seoul.

Later on Monday afternoon, Green and Lampman drew the assignment of destroying the embassy switchboard. This switchboard was composed of four eight-position manual banks which they went at with sledgehammers. It took a couple of hours to complete the job. Then came chow and more carpet time.

Lampman was awakened at 2200 and was told to get a radio jeep for a run to Kimpo. As the switchboard was out and the radio not secure, he had to personally contact the airport contractor, Bourne Associates, to remove all their heavy equipment from the runway - as FEAF would start running in planes as soon as dawn broke. The YAKs had shot out all the communications at Kimpo, and the last telephone instruction from the embassy had been to block the runway with heavy equipment to impede an airborne assault during the time the South Korean Army was still holding the North Koreans in the Ouijonbu corridor.

Lampman headed for Kimpo along with a Korean police lieutenant, an interpreter and a driver. The streets were quiet. They were challenged at almost every street corner through Seoul and Yong Dong Po and at each intersection in the dark countryside all the way to the airport. All the checkpoints were jittery. Lampman was jittery too, thinking about how it was in China in 1946 and '47 after curfew - the "halt" three times and "Bang" scenario. They made it to Kimpo, and the runway was cleared by first light. They departed Kimpo and arrived in Seoul as the sun came up, ate some chow and hit the carpet for much-needed sleep.

Throughout the early hours of Tuesday, 26 June, the document destruction continued and the remaining embassy staff was given breaks to pick up personal gear from their quarters. The limit was supposed to be one suitcase and what could be worn. This order produced some funny outfits: fur coats (in June), Easter bonnets, tennis rackets, hunting rifles, shotguns, ice skates, bottles of booze or perfume, bolts of silk, Irish linen, plus two or three cameras and extra purses hanging from the people's necks. The last remaining nonessential personnel were put in a bus convoy to leave the embassy for Kimpo Airfield at 0730. Again, the Marines led the way.

At Kimpo, passengers were loaded onto the transport aircraft in very fast order. Six planes were cycled through in forty-five minutes. The last plane out was attacked by North Korean YAK fighters. American P-51s and F-82s knocked down another North Korean plane, and the rest departed in a big hurry. There were no friendly losses. The convoy returned, mission accomplished, in the early afternoon. Final destruction duties were being carried out at the embassy, where only the Ambassador, consul, first secretary and the Marines remained. The Ambassador's group, with Jack Edwards as the single Marine guard, would be the last to depart - by road to the south.

Later on Tuesday morning Lampman was assigned a run out to the East Gate of the city to make sure the fuel depot supervisor was on his way in to the embassy. He was also to ensure the Seventh-day Adventists were aware of their own opportunity to fly out on Tuesday morning from Kimpo.

Lampman never reached the Seventh-day Adventist compound, as he sighted a North Korean tank on the road north of the East Gate. It was evidently out of fuel and had outstripped the infantry support. There didn't appear to be any organized friendly military force between the tank and Seoul. Had the North Korean armor been able to move, they could have captured the city. One distant look was enough, and Lampman went back into the city and to the embassy. It wasn't known at the time what happened to the Adventist group, but they did survive. They were seen later after the Marines returned to Seoul following the Inchon landing.

Lampman and Green's assignment at midmorning was to destroy two of the embassy code machines. With two or three armed escorts, they took the machines to the sidewalk in front of the embassy and hooked them up with leads to a jeep battery. The code machines were thermite encased, and upon starting the jeep engine the machines began to melt. In about three or four minutes all that was left were two lumps of molten metal, each about the size of a football.

Things became really hot then. Few people were left in the embassy, aside from the Ambassador's party. Dupras was assigned to take all the greenbacks out of the finance office safe. The finance officer half-filled a regular-sized U.S. Mail bag with packs of bills and told Dupras to "get this to Japan." That was it - no receipt, no lock, nothing.

Lampman was told to remove the Great Seal of the United States from the mechanism in the consulate and ensure it was destroyed. The Seal, which was used to certify official documents such as passports, was embossed on a square column of case-hardened steel about two inches square and three or four inches long. Having nothing capable of destroying it, all he could do was take it along on the evacuation in his jacket pocket. Since the Seal was not to be taken out of the country, Lampman would have to figure out some way to destroy it before getting to Japan.

Late Tuesday afternoon the Marines were told to pick out the best jeeps, service them, get some expeditionary cans of gas, gather some food and head south out of the city. The NCOIC, John Runck, who had spent a tough cruise in Japan as a prisoner of war, let it be known that none of the Marines were going to be captured, no matter what or how. When everybody else was out and headed south, the Marines too would be allowed to go. The word was that they should head for Pusan and when they got there - start swimming.

Dupras and Lampman found the assistant security officer's jeep, which had a radio. Then came a last-minute change of orders. The Marines were to head for Kimpo because FEAF was sending in one more plane. In late afternoon the Marines headed out for Kimpo in small groups as the escape vehicles were readied. The enemy was at the East Gate of the city.

As late as 1700 on the 27th, the South Korean chief of staff was still trying to build up a final defensive line north of Seoul. It was to no avail. One of the Ambassador's personal security guards, Jack Edwards - the only Marine not to come out through Kimpo - accompanied Ambassador Muccio and a small staff south by car, just ahead of the North Korean invaders.

The Marines navigated the streets of Seoul and went over the Han River Bridge, which was later prematurely blown up at 0215 on 28 June - trapping many of the South Korean forces still fighting a delaying action in the city. The road all the way through Yong Dong Po was heavy with refugees also heading south. Once on the Kimpo Highway, it was clear sailing. Arriving at Kimpo, the Marines found a large group of non-Korean nationals milling around. The Marines kept close to their vehicles on the tarmac - they still might need them - and waited and waited.

Finally one C-54 dropped in and taxied up to the tarmac in front of the terminal. By now Korean police had established order and John Stone, the American Consul General, cleared all of the people to get onto what was understood to be the last plane out of Kimpo. The pilot of that plane informed Stone that FEAF had one more "last plane" on the way in about one hour. That plane then took off, leaving everyone on the field just waiting. More and more people kept arriving, and soon what was to be the "real last plane" came in - and all those people wanted to get on it.

Everybody had been informed since early Monday that air evacuees could carry only one small bag. Needless to say all of the bags were big, and people also were carrying a grand assortment of loose gear and packages. Stone had a list of cleared people and also a big heart, so he piled people on the plane. The pilot never shut down his engines. The sound was deafening, and the plane was rocking on the runway.

At that point the only folks left on the ground were the Marines and a few embassy staff. The word came down from the pilot that he was already overloaded, but he said, "What the hell, come on!" So, the Marines too climbed aboard - standing room only forward. Contemplating the likely situation in Kimpo, FEAF had put on a minimum crew - pilot, copilot and crew chief. They had foreseen all the people who would want to board in Kimpo.

Some of the Marines wound up standing in the navigator's compartment. As soon as the last guy was aboard the door closed and the plane began to move down the taxiway toward the runway. Having put on the navigator's headset, Lampman heard the pilot say to the crew, "I don't know if I can get this thing off the ground." The pilot immediately told the crew chief to open the door and throw out anything that wasn't nailed down - and away the plane headed for the runway with a stream of suitcases, boxes, bags, survival gear and weapons littering the taxiway. The sound of machine-gun fire on the airfield perimeter was a real motivator. Almost everything went out of the plane. Little was sacred.

Everyone breathed easier as the plane burned off some fuel and gained altitude. There were 110 people aboard, which must have been some sort of record. Over the Straits of Tsushima, Lampman completed his final evacuation task. He received permission to open the little, round navigator's Plexiglas porthole and dropped the Great Seal of the United States into the briny depths of the Tsushima Strait.

Upon arrival in Inazuke, Japan everyone was warmly greeted by the large base reception committee in a hangar set up with refreshments, telephones and medical assistance. FEAF went all out and did a superb job. Everyone received a toilet kit, skivvies and socks. There was free food in any of the clubs, and there were bunks overnight before everyone boarded a train the next morning for Kyoto.

Upon landing in MacArthur's empire, anyone with greenbacks in their possession had to go to the base finance office to fill out a declaration and exchange the currency for Military Payment Currency - the legal scrip in Japan under the occupation. Dupras went to the Base Finance Office with the mailbag full of greenbacks. Since the possession of greenbacks in "MacArthurland" was equally as serious a crime as murder, it was easy to understand the commotion which ensued when Dupras dumped out the mailbag on the floor. However, Dupras had done his job well. Imagine if he had let the crew chief throw that bag out onto the Kimpo taxiway!

It had taken hard work throughout three eventful days to safely evacuate everyone, and it was accomplished without loss of life. The Ambassador said that it was done "without so much as a bloody nose."

A final note - Sunday through Tuesday, the majority of the Koreans employed by the American Embassy remained loyal and steadfast. The drivers, dining room staff, hotel employees, and switchboard and communications personnel all stayed on the job. It is doubtful whether the evacuation would have been so successful without their unstinting help. Unfortunately, they paid dearly for their loyalty. The North Koreans dealt severely with them and their families - those who were not able to go to cover during the occupation. Some were executed, and some were never seen again. They were the true heroes of the evacuation from Seoul.

RESOLUTE RESPONSE

MSG Detachment Nairobi, Kenya

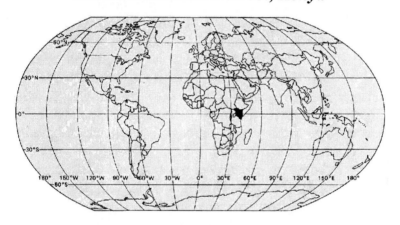

The Moscow Station scandal brought a lot of bad publicity to the Marine Security Guard program, but it was not a true indication of the type of Marines we had out there guarding our Embassies and Consulates. When the following incident occurred I took a bit of a personal interest since our Embassy in Nairobi was headquarters for Company F, and I had visited there on several occasions. These Marines are far more representative of the MSG program than the likes of Clayton Lonetree:

On the morning of August 7, 1998 Corporal Samuel Gonite was standing duty at Post One of the American Embassy in Nairobi. Another MSG who was off duty at the time, Sergeant Jesse Aliganga, stopped by to chat with Gonite for a few minutes on his way to cash some checks for a Marine social function scheduled for that evening. When their conversation concluded Aliganga left Post One and got on the elevator. It was at that moment that a bomb was detonated by terrorists in the Embassy parking garage.

Gonite heard the explosion and checked the closed circuit television monitors to see if there had been a car accident on the street in front of the Embassy. A split second later he was knocked to the ground by a second explosion. When he opened his eyes he couldn't see anything due to the dust, and a few moments later he heard someone pounding on the door to his post. He opened it and discovered it was his Detachment Commander, Gunnery Sergeant Gary Cross. Cross had been thrown to the bottom of a stairwell as he responded to the first explosion, and he directed Gonite to secure classified material and aid people as they evacuated.

At the time of the explosion two other Marines, Sergeants Aaron Russell and Daniel Briehl, were waiting for Alingala by their car in front of the Embassy. They decided to go inside to wait, and had just started moving toward the entrance when the bomb exploded. Briehl dove for cover by their car as debris began to fall on him, and Russell sprinted to Post One to check on Gonite. A few moments later Russell heard a muffled yell and realized that Briehl, who had followed him into the Embassy by that time, had fallen down the elevator shaft while trying to feel his way through the smoke to Post One. Debris was falling down the shaft on top of Briehl, who had sustained three broken ribs in the fall. Despite his injuries, he assisted a group of people who had been trapped below in climbing up and out to safety. When Cross found him a bit later he sent the Marine to the hospital for treatment.

Gunny Cross immediately set about the task of securing a perimeter around the Embassy. He, Russell and Gonite donned their react gear and surrounded the Embassy as best they could until Cross was able to enlist the aid of additional military personnel.

While all this was happening Sergeants Armando Jiminez and Raymond Outt heard what was happening over their radio at the Marine House. They ran outside, flagged down a passing diplomatic vehicle, and headed for the Embassy. When they couldn't get through the snarl of traffic the Marines jumped from the vehicle and ran to the Embassy. Their Detachment Commander later commented that they "probably ran the fastest PFT of their lives." Upon arriving at the scene thirty minutes after the explosion the two reported to Cross and asked what they could do.

Another member of the detachment, Sergeant Harper, had been on leave in Mombasa where he had gone on safari. As he returned from the airport later that evening he heard about the bombing, and he headed straight for the Embassy.

Eventually seven search parties were organized, and Gunny Cross was asked if his Marines would be willing to help since they were intimately familiar with the layout of the Embassy. It was emphasized that since the building was unstable it was strictly a voluntary effort, and no one could be ordered to go back inside. Every single Marine volunteered without hesitation.

Cross and his detachment immediately reentered the shattered building in an effort to locate Aliganga and other Embassy personnel who had been killed or injured. Digging with their bare hands at first, they were determined to locate their brother Marine. Twenty-seven hours later they found him – dead.

The Marines draped their comrade in an Embassy flag that had been blown from its pole by the blast, and when Cross, Outt, Harper and Jiminez carried out the body onlookers grew silent.

Guarding what was left of American Embassy Nairobi, Kenya.

At one point, as Cross helped carry a stretcher out of the Embassy, he noticed that Briehl was back from the hospital and standing guard. He was still wearing his hospital gown, over which he had donned his flak jacket and helmet. When Cross asked him what he was doing there Briehl simply replied, "I knew you needed me down here, so I checked myself out of the hospital." Cross ordered him to go back.

The six Marines had been on duty for roughly thirty-six hours by the time MSGs from other posts in the region arrived to relieve them. Search and recovery efforts continued for the next six days, and the Marines continued their mission. A senior State Department official commented that "the Marines did a tremendous job. While most people were running out of the building, they were running in, despite the obvious danger."

Many of the Embassy staff described the building as

unrecognizable after the bombing, but through the smoke, flames, death and destruction one thing remained easy to recognize. Esprit de Corps.

ABODE OF PEACE

MSG Detachment Dar es Salaam, Tanzania

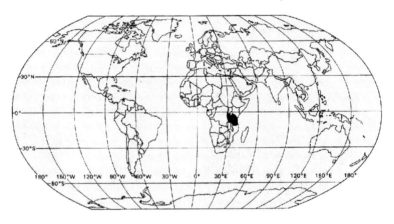

In case you didn't know, Dar es Salaam is Arabic for "Abode of Peace." How ironic. The day after the Nairobi bombing Islamic terrorists – who know all too well what the name means – bombed our embassy there.

On Friday morning, 8 August 1998 the Regional Security Officer for the American Embassy in Dar es Salaam, Tanzania, John DiCarlo - a former Marine Security Guard himself - had taken Post One from Corporal Johnson in order for him to make a head call. At approximately 10:40 AM a truck bomb exploded outside the security gate, ripping through the embassy. Corporal Johnson was knocked to the deck. He stood up and immediately ran to Post One where he found it intact, but inoperable. Corporal Johnson immediately reacted the Marines, who were all at the Marine BEQ during the explosion, and informed all mission

132

personnel to evacuate the building. He then donned his react gear and took control of the Command Center. The Detachment Commander, Gunnery Sergeant Kimble, arrived at the Embassy approximately four to five minutes after the blast and began checking offices throughout the chancery to ensure all personnel were safely out of the building. One of the casualties of the explosion was Gunnery Sergeant Kimble's wife Cynthia, who sustained bruises and eye injuries from flying glass – but while Cynthia was flown to London to receive eye surgery, Gunnery Sergeant Kimble never lost focus on the mission at hand.

American Embassy Dar es Salaam after the explosion.

Within eight minutes of the blast Sergeant Sivason and "first poster" Corporals Bohn, Hatfield, Johnson, and McCabe began working through the chancery clearing all rooms of personnel. No direction was required, as each Marine knew exactly what had to be done. Keep in mind that numerous secondary explosions (which turned out to be the fuel tanks of automobiles) were occurring all around the compound as the Marines continued to sweep the building. For a period of time, they had no idea if the embassy was actually being assaulted or overrun. Due to the enormous amount of smoke and fire raging throughout the chancery and near the underground fuel tanks, the decision was made to evacuate Post One and the Marines fell back to their secondary positions.

Corporal Johnson provided security for the mission personnel at the rear of the Embassy, and all other Marines took up perimeter security around the building. The force of the blast blew out every window in the chancery, and all doors except for Post One. The hard-line doors, which are located on the opposite side of the embassy, were also forced open by the blast.

Emergency fire exits on the opposite side of the building from which the blast occurred were blown off the hinges. Concrete walls within the Embassy were knocked down and safes were moved, and in some cases knocked over. During the search of the building the Marines had to bust through walls in order to get to areas unreachable during their sweep. Within four hours of the truck bomb, which damaged diplomatic properties and houses up to one thousand meters away, the embassy was secure with MSGs maintaining twenty-four hour security on the building until the arrival of the FAST team.

During a meeting called by the Charge d' Affairs in Dar es Salaam just days after the attack to personally recognize the Marines for their heroism, the following statement was made, "The Marines are to be commended for how exceptionally well they performed their duties under extreme conditions of chaos and terror. Their bravery and heroism was displayed in such a confident and purposeful manner that their very presence transferred to others, allowing them to get through the situation."

VINTAGE CAIRO

MSG Detachment Cairo, Egypt

Ron Harwood

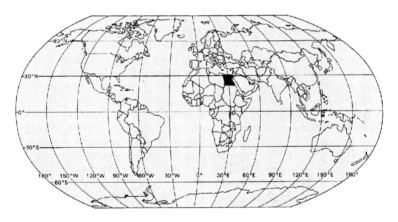

A lot has changed in Cairo since this took place back in the 1950s, but even so it is a good example of things which occur at diplomatic posts around the world – and much more often than you might imagine.

I was stationed at the American Embassy in Cairo, Egypt from October 1951 to February 1954. During that time we witnessed riots, the overthrow of a government, and also found time to visit many of the historical sites in the "cradle of civilization." The day we arrived (three replacements - Sergeants McQueen, Greer and Harwood) we were met at the airport by one of the Marines we were replacing. He immediately informed us that we would have to travel through Cairo to the Embassy by a round-about route as there was the distinct possibility we would meet rioters. We did wonder what kind of a greeting this was, but climbed in the jeep and headed out. Needless to say, we made it without real incident.

The first few months were relatively quiet although we did not venture out alone. One thing in our favor was the Egyptians really liked Americans at that time and we were known by our wide, rather loud, neckties. On 3 January, 1952 I met my future wife on the steps of Shepherd's Hotel, where "the east meets the west" and we enjoyed "SB's" (the specialty of the bar) on the terrace.

Serving in the shadow of the Great Pyramid can have its moments.

Three weeks later the hotel was no more, burned beyond repair on "black Saturday" when the natives rioted in earnest. Fires started up all over Cairo, seemingly at the same time. We were able to see into one of the main squares

137

of the city from the roof of the USIA Building on the Embassy compound, and watched as places with signs in English were torched while police and firemen stood by and watched.

We were soon sent to our assigned posts in preparation for defending the Embassy if it was a target. A few rioters reached the front gate but were turned away by Staff Sergeant Albers, who was a rather imposing figure armed with a tommy gun. By the way - some members of the embassy staff did not even know we had weapons prior to that time.

By mid-afternoon things were no better, but we did not appear to be in any immediate danger. I was called to the office of the NCOIC and told to take a message from our Ambassador to the Egyptian Ministry of the Interior, which I proceeded to do via an embassy staff car and driver. Later I was to learn that the message had words to the effect, "stop the rioting or you will be visited by Marines from the Sixth Fleet." Soon after that trip, the Egyptian Army moved and things calmed down rather quickly.

In the end, the reign of King Farouk ended and General Naguib led the country for about a year until Nassar came to power. The rest of my tour was uneventful for the most part, however I did receive permission to get married and remain on embassy duty. We were the first to receive that permission from both the State Department and HQMC. My wife Nell was teaching school at a Presbyterian Missionary School in Assiut, Egypt when we met, and we are still together and enjoying retirement.

THEY PLAYED
White Christmas

MSG Detachment Saigon, South Vietnam

R. R. Keene

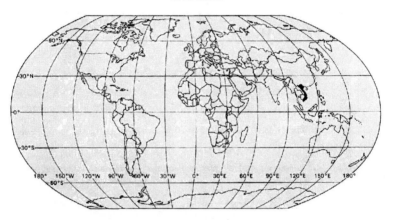

The heroic deeds of the Marines guarding our Embassy in Saigon often go overlooked – probably because a lot of people didn't even realize there was an MSG Detachment there during the Vietnam War because such things were overshadowed by our large military presence in that country. I imagine things are now much the same for the detachment in Baghdad – although I hope things turn out better for them in the end!

As if conjured by the farsighted imagination of a Greek tragedian, the final days of the Vietnam War ended in bitter paradox. America's noble ambition at the war's beginning - to champion democracy and aid a people menaced by

139

communist aggression - had gradually spiraled into disillusionment and ignominy. This sadness which President Ford painfully described, the final brush stroke to a peculiar masterpiece ten thousand days in the making, intimately involved men whose duty it was to protect and defend the American Embassy in Saigon. This burden, arguably the darkest hour in American military history, was shouldered by a special breed and remains a significant yet overlooked event in Marine lore. As enemy tanks rumbled into Saigon, the last vestige of U.S. military presence in Vietnam was lifted via helicopter from the embassy rooftop in April of 1975. Manning walls much like those individuals immortalized at the Alamo did, these defenders went sleepless and hungry for days, saving countless lives during an interval filled with chaos and hysteria. This is their story, their insights and reflections, of the Marine Security Guards of Saigon.

Vietnamese civilians coming over the Embassy wall as Saigon falls.

In the early hours of 29 April 1975, the grim and undeniable reality became apparent to the highest-ranking American official in Vietnam. For weeks, an uneasy tension had mounted in Saigon as the North Vietnamese Army began an aggressive and largely unabated sweep down the coast of the South China Sea. Da Nang had fallen less than a month prior, prompting a panic-stricken exodus of South Vietnamese soldiers and civilians alike. Two weeks before that, in neighboring Cambodia, Marines and 7th Fleet sailors evacuated U.S. personnel from Phnom Penh as communist Khmer Rouge forces began to overrun the capital. The final South Vietnamese resistance was overwhelmed by three NVA divisions on 20 April at Xuan Loc, located only 38 miles northwest of the capital city. As both South Vietnamese defense and spirit crumbled, President Nguyen Van Thieu transferred power on 21 April to ailing Vice President Tran Van Huong before the National Assembly. Hanoi's minister of defense - and the mastermind of the French defeat at Dien Bien Phu thirty years earlier - sensed this was the long-awaited sign that victory was at hand. Quickly seizing the momentum, General Vo Nguyen Giap ordered an all-out assault on the southern capital. State Department officials, warily monitoring the events from Washington, D.C., began to realize the situation was untenable. Scores of NVA rockets and artillery shells began to pound Tan Son Nhut air base. Thousands of desperate Vietnamese were besieging the embassy with hopes that either through bribery, sympathy or luck, they too might accompany the retreating Americans. Graham Martin, American Ambassador to South Vietnam, finally received the call he dreaded: President Ford had approved and directed Option IV, the helicopter evacuation of Saigon. Operation *Frequent Wind* officially began shortly before

1100 on 29 April, when Armed Forces Radio broadcasted *"(I'm Dreaming of a) White Christmas."*

The miracle for which Martin waited - a heroic, last-ditch defense at Xuan Loc or perhaps a last-minute negotiation with the North Vietnamese to avoid invasion of the city - never materialized. His confidence now rested on the Marine Security Guards who manned the embassy walls, the Marine Aircraft Group 36 helicopter pilots who would execute the mission, and Marines and sailors aboard 7th Fleet ships located in the South China Sea just over the horizon.

Daily flights began evacuating up to five hundred U.S. personnel, foreign nationals and "at-risk" Vietnamese - those who supported the U.S. government - in early April when the South Vietnamese collapse loomed. The key to these large-scale evacuations was Tan Son Nhut, which served as the Defense Attaché Office (DAO) command center and departure point for large, fixed-wing aircraft. As the state of affairs began to deteriorate, sixteen of the forty-five Saigon MSGs were siphoned off to assist in processing and providing security at the DAO on 19 April. "I didn't like the idea of splitting my forces," recalled then-Master Sergeant Juan J. Valdez, "but we were under the operational control of the State Department, and what they said was it."

Like many of the senior leaders in Saigon at the time, Valdez was well-seasoned and had seen many swings of the pendulum in Vietnam. During the early stages of the war the San Antonio native served a two-year tour from 1965 to 1967 with Company B, 3d Amphibian Tractor Battalion, attached to 2nd Bn, Fourth Marine Regiment. He returned once again in September 1974, this time as the Saigon detachment noncommissioned officer in charge. Initial embassy estimates predicted that approximately seven thousand Americans would seek safe passage out of Saigon.

"It now seemed virtually impossible to estimate how many Americans were living in Saigon and nearby Bien Hoa," said Valdez. A sense of uncertainty intensified daily as NVA forces gradually tightened the noose around the city. "A Vietnamese marriage certificate, which only a few months before had cost no more than twenty dollars, now cost up to two thousand," said Valdez. "The crowds never appeared dangerous, just desperate - begging [to leave] the country or get their children off to safety." In many regards, the situation facing the MSGs was a prototype for the modern battlefield Marines are predicted to inhabit: an uncertain, chaotic arena where the lines between open conflict, humanitarian assistance and peacekeeping are blurry at best. How many wolves are among the sheep? Do sheep left in the wilderness transform into wolves? Valdez was relying upon many young, inexperienced Marines to act decisively in matters of life and death - perhaps their own, and undoubtedly on behalf of others. Bill English, one of the young MSGs assigned to the DAO, reported to the Saigon detachment shortly before the evacuations began. After checking into the Marine House, the "new guy" remembered trudging up a long flight of stairs, selecting a room and looking out over Saigon, "trying to figure out how I had gotten here and what I was going to see in the coming days." A lance corporal with seven months on active duty, English suddenly found himself roaming the compound on night watches. What was once the DAO movie theater had evolved into an evacuee processing center. As his footsteps echoed throughout the gymnasium, a staging area where nervous hopefuls awaited their freedom bird, "One gentleman came out to the hall and told me how comforting it was to hear, like an affirmation of our presence," he said. "I realized that these anxious people took comfort out of the rhythmic sound of our marching in the halls." In a 1995 *Time* magazine article, one of the architects of the Saigon assault, NVA Lieutenant General Hoan Phuong, described how the final nails were to be thrust into the South's spirit. His army "enlisted" South Vietnamese air force pilots who were primarily driven to curry favor with the conquering army, but who partially wanted to strike against the Americans abandoning them. Employing South Vietnamese Air Force, American-produced, A-37 Dragonfly jets and F-5 Tiger aircraft, the defecting pilots were ordered to strike key locations in Saigon. "The idea was to bomb the concrete hangars and the runways at Tan Son Nhut," Phuong recounted. "We didn't think we'd do much real damage, but we wanted to have maximum psychological effect. We wanted to create chaos." The chaos Phuong's attacks created - both psychological and tangible - radically altered both the timeline and the strategy in which U.S. leadership attempted to evacuate its personnel. At approximately 1630 on 28 April the defecting pilots attacked Tan Son Nhut air base, targeting the DAO command center and control tower.

Although a forty-member supplementary platoon composed of 9th Marine Amphibious Brigade Marines from Okinawa had arrived a few days earlier to help provide security, the MSGs constantly manned the compound's primary positions. Post Two was located at the intersection of the main road into the airport, and Post One was positioned thirty yards away at the road leading into the DAO compound. As the dust from the earlier attack began to settle, it was time to begin assigning guard watches and dig in for what was expected to be an uneventful evening. "I stayed up that night and, at around 0200 on the 29th, walked around to check on the Marines at their posts," said Sergeant Ted Murray, who had arrived at the Saigon detachment the previous December. "Almost all of them smiled and asked for some sleep, something that we all needed, but they were Marines - embassy Marines - and they knew their job." The last MSGs who Murray visited were Lance Corporal Darwin Judge and Corporal Charles McMahon Jr., who assumed their Post One positions at midnight. At approximately 0330 a series of randomly launched rockets began pounding the air base. As Marines poured from their quarters and began to assess the situation it was discovered that Post One, the closest position to the main gate, had taken a direct hit. The two MSGs manning this position, Judge and McMahon, had been killed during the attack, and subsequently became the last U.S. service members to die as a result of enemy fire on Vietnamese soil.

Since arriving at the Defense Attaché Office on 16 April 1975, Marine security guards Judge and McMahon were primarily responsible for assisting evacuees during processing and manning security posts. A steady stream of American, Vietnamese and foreign national evacuees had passed through the DAO compound, but as the advancing North Vietnamese Army gradually tightened the noose around Saigon, the pressure was beginning to mount.

Sergeant Doug Potratz and his family were among the multitudes seeking safe passage to American soil. Throughout his last month in-country, Potratz displayed an unerring knack for making crucial decisions on particularly ominous occasions. He married his Vietnamese girlfriend on 4 April - the same day Da Nang fell to the communists. He then arrived at Tan Son Nhut air base with his wife and four-year-old stepdaughter the same day South Vietnamese President Nguyen Van Thieu resigned from office, 21 April.

Frustrated by red tape, and after endless hours of waiting and fruitless attempts at securing a flight out of the country,

"I was ready to scream," Potratz recalled. "Judge came up to me and said, 'Sergeant Potratz, I know the guy who fills out the plane manifest. Give me your paperwork, and I'll get your family on the next flight out.'"

Displaying typical Marine resourcefulness, Judge returned a few minutes later, picked up Potratz's stepdaughter and a suitcase, and escorted the family to the plane. "That was the last time I saw Darwin Judge alive," Potratz said. "He was my hero that day."

The days and hours leading up to 29 April were becoming increasingly tense, and as one MSG described "full of action, boredom and turmoil." Responsible for posting the guard that night was Sergeant Kevin Maloney who, like McMahon, spoke with a thick Bostonian accent. The two Massachusetts natives were originally scheduled for the midnight watch at Post One - a position at the DAO compound's outer gate - but buddies Judge and McMahon requested to be posted together. "I reasoned that no real action would occur until morning [and that] I should be where the action was," said Maloney.

At midnight McMahon and Judge relieved Bill English who, like a somnambulist, trudged to his rack and settled down for a well-deserved rest. Less than four hours later the base came under attack by North Vietnamese rockets launched from nearby positions. Grabbing their weapons and gear, English and his fellow Marines scrambled to reach bunkers located outside the building. They soon discovered that Post One had taken a direct hit, and both McMahon and Judge had been killed.

Because Judge and McMahon exemplified the Marine spirit - exhibiting compassion and professionalism during a bleak, extremely confusing period - they remain both admired and honored by the MSGs who served in Saigon.

One man who can testify to this is Potratz, who still remembers the actions of a young lance corporal on his behalf.

"If it weren't for the 'Darwin Judges' and the 'Charles McMahon's,'" he reflected, "thousands of Americans and Vietnamese would not have made it out of the country and lived a fuller life."

As the sounds of artillery, rockets and gunfire echoed throughout the city, Ambassador Martin wanted to personally inspect the damage inflicted upon the air base. Martin called upon Master Sergeant Colin Broussard and Staff Sergeant James Daisey, two of the six-man Personal Protective Security Unit assigned to him on 29 April. "The streets were lined with Vietnamese," recalled Broussard, who escorted the Ambassador the six treacherous miles to the DAO. "We didn't know who was the enemy. We locked and loaded all weapons and could almost feel an attack on the motorcade."

Awaiting the Ambassador was a smoking, pockmarked air base in flames and still receiving sporadic rocket fire. Acknowledging that fixed-wing evacuations from the air base were no longer viable, Broussard said, "The Ambassador saw what he wanted to see and ordered us to bring him back to the embassy." It was becoming increasingly apparent to Saigon's residents that the end was near. As the NVA began its thrust into the city Vietnamese throngs began seeking any feasible way to reach the U.S. Navy flotilla, Task Force 76, positioned twenty miles offshore. South Vietnamese soldiers commandeered military aircraft, and civilians flocked to all ports along the Saigon River in hopes of reaching the Americans at sea. Around 1100, the same time *White Christmas* began filling Armed Forces Radio airwaves, Martin requested a security squad to escort him to his residence, located about two blocks from the embassy. Reports of Viet Cong assassination squads, snipers and continual rocket fire failed to dissuade the Ambassador, and Broussard and Daisey were once again among those called upon for a dangerous assignment. "We brought Uzis, grenades and .357s with us and went through a secret entrance in the French Embassy," said Broussard. "We went into the house and burned classified information and used thermite grenades to destroy sensitive items."

In the interim, 9th MAB units aboard Task Force 76 ships were gearing up for yet another historic evacuation. The first wave of Marine Heavy Helicopter Squadron 462 aircraft, loaded with the 2nd Battalion Landing Team of the 4th Marines, touched down on DAO landing zones at approximately 1500. The reinforcements rushed to their assigned positions as evacuees began boarding the initial twelve Sea Stallions. During the ensuing nine hours, 395 Americans and nearly 4,500 Vietnamese and foreign nationals were airlifted from the DAO to waiting ships. The last elements of BLT 2/4 lifted from the DAO just before midnight, concluding what had been an orderly, well-executed evacuation. The situation at the embassy, however, was much more volatile as drastically outnumbered Marines attempted to keep at bay approximately ten thousand frantic Vietnamese surrounding the embassy walls. The front gate had been secured in order to keep the human tidal wave from flowing onto the embassy grounds, and the MSGs were finding it extremely difficult to assist those marked for evacuation. "If there was someone out there that we wanted to bring in," explained Major Jim Kean, the Saigon detachment officer in charge, "then we'd put a bunch of people on the wall, reach down, grab him by the collar and hair and just yank him up and over the wall." The advancing

North Vietnamese didn't interfere with the evacuation, but helicopter pilots received small-arms fire while hovering over the embassy. This "cowboy-style" shooting by South Vietnamese rogue looters further complicated an already perilous task. While CH-46s landed on the embassy roof, pilots of the larger CH-53s were forced to execute a steep descent into the embassy courtyard. "There had been waves of choppers. One in the air hovering, and one on the ground loading," Valdez recalled. The Vietnamese inside the embassy were assured that no one was going to be left behind. "We had to run around counting people to see who was going to get out and who was not going to get out. It was grim," said Kean. "At any time during the night, the number of people inside the grounds seemed to remain steady." Evacuations began late that afternoon and continued steadily throughout the evening, but the fate of those remaining inside the embassy walls was finally sealed when a helicopter with the call sign "Lady Ace 09" touched down on the embassy roof shortly after 0530 on 30 April. The pilot had received specific orders that he was to extract the Ambassador and his staff, and that all further flights were designated strictly for U.S. personnel. "Martin looked over at me for a moment. He didn't say anything, and he didn't show any emotion. He just looked tired," Kean described. "He knew that this sad moment would be coming sooner or later. Then he went upstairs and got in the bird and left Vietnam. He was carrying an American flag with him." As the Marines began to withdraw from the perimeter, cautiously backing toward the embassy door in an effort to "button up" inside, Vietnamese who had been promised liberation suddenly realized they were about to be left behind. The Marines barricaded the doors, froze the two elevators on the sixth floor, and made their way to the rooftop landing zone.

"Then everything came to a standstill and we just sat," said Valdez. "All the Marines were up there. No birds in sight. But I never thought for one minute that the choppers would leave us behind." Marine pilots accumulated 1,054 flight hours and flew 682 sorties throughout Operation Frequent Wind, evacuating five thousand from Tan Son Nhut and more than two thousand from the embassy. At its apex, America's military presence in Vietnam had numbered half a million. It was now reduced to BLT 2/4 reinforcements and the MSGs - sixty isolated Marines on a rooftop overlooking a city under siege. One by one, the final series of helicopters touched down and evacuated the infantry Marines, until eleven MSGs were all that remained. "Some time just before 8 AM I saw the bird off in the distance - one unescorted CH-46 out of the sunrise," said Kean. The largest helicopter evacuation in history, as well as America's twenty-five-year struggle to keep South Vietnam free, ended a few moments later as the last Marine CH-46 lifted from the embassy and headed out to sea.

SAIGON BODYGUARDS

MSG Detachment Saigon, South Vietnam

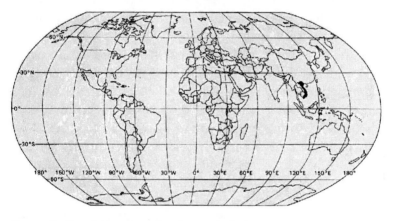

One of the duties rarely discussed in conversations about MSG duty is the possibility of being assigned to the Personal Protection Detail of a VIP. As anyone who has ever drawn such an assignment knows – there's never a dull moment!

United States Ambassador to South Vietnam Graham Martin sat in the darkened living room of his residence with his Marine bodyguards watching *Hitler's Last Days*, a saga of the Furher's deathwatch in the Berlin bunker. As the lights came up after the first reel, Martin sighed, "You know, I'm beginning to realize what that SOB felt like." The young Marines of his Personnel Protective Security Unit (PPSU) looked knowingly at one another. South Vietnam was disintegrating and thousands of victorious North Vietnamese Army (NVA) troops were advancing on Saigon. It was only a matter of days until they reached the capital. NVA death

151

squads were already loose in the city, randomly attacking government facilities and threatening the security and safety of the Embassy staff. Apprehension was high among the six members of the PPSU - the life of the Ambassador was in their hands.

Staff Sergeant Dwight McDonald was detailed to provide security for Ambassador while he attended a South Vietnamese Boy Scout Jamboree. "We were at war, with almost no security, and the Ambassador wants to go out in the country with President Thieu. My NCOIC, Staff Sergeant Clem Segura, told me he didn't want to see any daylight between the Ambassador and me. After the ceremony the Ambassador stood in a reception line, with me close at his side. President Thieu came along and, since I'm between him and the Ambassador, he had to shake my hand - making him mad as hell. Tough - no one was going to get this Marine's Ambassador!"

The six-man Saigon PPSU were specially trained Marines assigned to the American Ambassador. Retired Master Gunnery Sergeant Colin Broussard (then Staff Sergeant) recalled, "After graduation from Security Guard school I was accepted into the classified six month Bodyguard School with the State Department where I trained with various government agencies. Our training in Vietnam wasn't classified. We were required to learn every main street and safe haven in the city, memorize code lists, and be proficient with our weapons - Swedish K and Uzi submachine guns, .357 magnum pistol, and the M39 grenade launcher." The PPSU men wore civilian clothes and according to Broussard, "(We) turned in our military ID cards for Vietnamese credentials that identified us as civilians with a GS-9 rank, the equivalent to an ARVN Lieutenant Colonel."

The PPSU house was located adjacent to the

Ambassador's residence on Phan Quac Quan Street, about three blocks from the Embassy. It had four bedrooms, an armory, kitchen and a communication room filled with sophisticated equipment. A door led directly into the residence grounds, which were surrounded by concrete walls eight-feet high. Two metal towers on each corner provided overlook positions. Embassy Marines walked the perimeter during the day but, as Broussard remembered, "We (PPSU) took over security of the residence at night because we thought if there was an attack, it would be after dark. In late February/early March all non-mission critical Embassy employees were flown out of the country and the city reeled under NVA sapper attacks. We were getting more and more anxious."

On 10 March, 1975 the North Vietnamese launched their final offensive against the South, committing over a 150,000 men in a bid to forcefully reunify the country. The South Vietnamese committed a devastating series of tactical blunders, which allowed the NVA to overrun two thirds of the country in a matter of weeks. The social and military fabric of South Vietnam disintegrated. Thousands of panic-stricken civilians and the tattered remnants of military units clogged the roads, desperate to escape the advancing Bo Dai - NVA infantry. Broussard remembered, "The ARVN (South Vietnamese Army) deserted in droves after the fall of Hue City and DaNang, starting a mass exodus from the north." Hue, with its famous Imperial Citadel, had been the site of a horrific battle during the 1968 Tet offensive when the NVA infiltrated the city. Its liberation had cost the lives of hundreds of U.S. Marines and soldiers, as well as thousands of South Vietnamese soldiers and civilians. Just before the city fell Broussard and another PPSU Marine, Staff Sergeant Jim Daisey, accompanied the Consul General of DaNang on a quick fact-finding trip. "The three of us flew to Hue in an Air America helicopter," Broussard recalled, "and landed on the south side of the Perfume River, across from the Citadel, near a house flying a large South Vietnamese flag. We stood guard while the Consul General talked to one of the government officials. The sounds of battle were extremely loud - artillery, small arms fire, machine guns - as the ARVN made their last stand." "We flew out two hours later, just ahead of the NVA!" Further south DaNang, the former headquarters of the 3rd Marine Amphibious Force, fell without a fight. The NVA captured tons of weapons and equipment, including flyable helicopters and jet aircraft. Four incredible weeks later the end was in sight – and the North Vietnamese commenced the final assault on Saigon's weak defenses.

In the early morning hours of 29 April a 122mm rocket slammed into Tan Son Nhut airport, killing two Marine sentries - Corporal Charles McMahon and Lance Corporal Darwin Judge - and wounding Corporal Otis Holmes. Sergeant Kevin Maloney of the PPSU was the first man on the scene. "Post One took a direct hit, instantly killing both men. A Vietnamese ambulance crew picked up the two remains, but I was little help to them. The sight of the shattered bodies had terrorized me." Heavy shelling continued, threatening to close the runway and prevent fixed wing aircraft from evacuating hundreds of Vietnamese. Ambassador Martin refused to believe reports that the airstrip was unusable and decided to personally inspect it. Broussard, the Deputy Chief of Mission and other Embassy officials tried to talk Ambassador Martin out of it, but he was determined to go. "He ordered me to get his car - a black armor-plated Chevy with a 450 cubic inch engine. Segura and McDonald took off in a jeep to find a secure route. Daisy drove the Chevy with Sergeant Paul Gozgit as escort. Maloney and I were in the follow-up car, along with two South Vietnamese Special Police officers. All of us were armed to the teeth and locked and loaded!" The streets were lined with hundreds of Vietnamese as the caravan sped to the airport. Broussard said, "I pointed my Swedish K out the window, with my finger on the trigger, thinking an attack would happen any second."

The motorcade arrived at the airport just after a heavy shelling. According to Broussard, "We got to the gate and the ARVN sentries wouldn't let us in. We jumped out and surrounded the Ambassador's car until McDonald, the big Missourian, 'talked' the guards into opening the gate. Plumes of deep black and gray smoke marked the funeral pyre of burning aircraft. Debris covered the shell-cratered runway." Martin viewed the devastation and made up his mind. Minutes later the convoy made its way back to the embassy at seventy miles per hour. Hundreds of frightened Vietnamese now surrounded the Embassy, clamoring to get inside. Some of them pounded on the cars as they entered the rear gate. At a little before 11 AM Martin called the Secretary of State, Henry Kissinger, and told him it was time for Frequent Wind's "Option 4," the helicopter evacuation of the Embassy.

As they waited for the evacuation to start, Martin decided to go back to his residence. His staff protested without success, and Broussard was ordered to take him in the car. "Jim (Daisey) and I shielded the Ambassador from the crowd as he climbed into the car. The gate was opened and I tried to drive through, but there were too many Vietnamese and they almost overran us. Several Embassy Marines came to our rescue as I backed up, while they fought to close the gate." Martin was undeterred and ordered his two remaining bodyguards - the other four were out collecting the families of the Vietnamese Special Police - to accompany him on foot. "We're going for a little walk," he told them. Broussard said, "Jim and I looked at each other, thinking this was the end. We grabbed our Swedish Ks, grenades, and .357s and led the Ambassador through a secret entrance into the grounds of the French Embassy and out into the street - the pucker factor was high!" The two bodyguards, with Martin "sandwiched" in between them, made it through the first street without incident. However, as they crossed the street, "Two Vietnamese Cowboys (kids carrying M1 Carbines) on a motorcycle stopped us," Broussard recalled. "Jim pointed his submachine gun at them and they took off." The three men made their way into the residence, even as a terrific firefight exploded in the cemetery across the street. The two Marine Guards let them in. Broussard added, "I started a fire in the house to burn secret documents that the Ambassador kept handing to me, while Jim went next door to the PPSU house and destroyed the classified radio/telephone equipment with thermite grenades." Unwilling to risk the walk back, they decided to use the backup car - an armored Pontiac. "I tried to start it, but the high compression engine wouldn't turn over," Broussard remembered. "I tried several times and finally it roared to life. Jim got the Ambassador in the back seat and laid over him, while the other two Marines jumped in and pointed their M-16s out the window. I floored it and rammed the eight foot high gates, which gave way under the impact of the heavy car." Broussard drove to the French Embassy, after deciding they couldn't get through to their own compound. The French Ambassador met them, and after exchanging remarks with Martin handed him a going away gift - a foot tall statue. Martin, in turn, handed it to one of his escorts, who had a devil of time trying to carry it and his rifle at the same time. He solved the problem by giving up his weapon. After ducking through the concealed door, they escorted Martin back to his office. Broussard remembers, "The Ambassador looked at us and said, 'Boys, I owe you a fifth of Scotch!'"

Just after 1600 the first evacuation helicopters settled into the compound. Broussard observed, "Marine infantry ran out the back and quickly manned the walls, reinforcing the exhausted Embassy Guard. It was an emotional moment, as we were feeling awfully lonely." The compound was a scene of chaos - hundreds of Vietnamese and Americans milled around, abandoned luggage and discarded weapons littered the ground, shredded documents and papers swirled about - and overlaying everything was the deafening noise of descending helicopters, scattered small arms fire, and exploding rocket and artillery fire. Outside, panic-stricken Vietnamese tried to scale the walls, even as Marines pushed them back. A woman threw her baby over the wall, and a man handed one of the Marines a bag filled with precious jewels and gestured to get in – but the Marine shook his head and handed the valuables back. Maloney remembered, "The crowd clawed at us as we pushed and shoved. We gave up the wall and retreated into the Chancellery. Marines destroyed communication equipment, while others kept the Vietnamese at bay with tear gas. We worked our way up, floor by floor, to the rooftop pad."

Marine Ch-46 helicopters continued the evacuation. Maloney said, "My turn to leave came in the morning twilight. I could see movement in the streets below, but streams of tracers held my attention. As we flew over the South China Sea, I could see thousands of surface craft of all descriptions, singularly and in small groups, heading toward our fleet."

Broussard added, "Around 0300 the Ambassador ordered Daisey and I to get on the next helicopter. We wanted to fly out with him, but an order was an order, so we made our way up the staircase with some of the infantry Marines and caught the next flight." A few minutes later, Ambassador Martin and his two remaining bodyguards, Segura and McDonald, boarded Lady Ace 09 and flew to *USS Blueridge*. The last contingent of Marines was pulled from the rooftop about an hour before an NVA T-54 tank pulled up in front of the Embassy.

Broussard's CH-46 landed on the flight deck of *USS Midway*. He and Daisey, numb with fatigue, staggered off the ramp into the arms of a Marine First Sergeant, who tried to grab the hand grenades off Broussard's flak jacket. The exhausted bodyguard pushed him away, walked to the edge of the flight deck and tossed his Swedish K, .357 magnum pistol and ammunition into the South China Sea, and then found an empty space in the passageway and fell into a deep sleep - his war was over.

The next day the bodyguards were reunited with the Ambassador on the command ship, *USS Blueridge*, but parted company in the Philippines. Ambassador Martin never honored his promise of the fifth of Scotch.

TERROR IN TEHRAN

MSG Detachment Tehran, Iran

Sergeant Stewart D. Hill

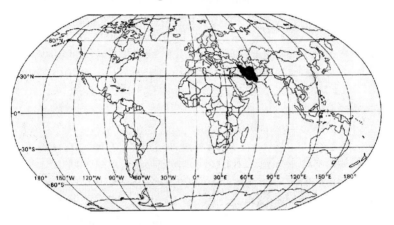

How MSGs are employed is left to the discretion of the Chief of Mission, which can sometimes be a problem because they are by nature always looking for the "diplomatic" way out of every difficulty. Marines, on the other hand, are men of action who look at things from a tactical perspective – but by the time the diplomats come around to their way of thinking it's often too late to take aggressive measures.

I learned very quickly the definition of a "hardship post" when my first assignment after MSG school was at the U.S. Embassy in Tehran, Iran. I arrived in Tehran on 7 December 1978, and although the Revolution had not started yet tensions in the city were high. Martial law was in effect, and a curfew was set at 2100 hrs. Any person on the streets after

159

2100 was subject to a confrontation with the Iranian army who patrolled the streets.

Myself and six other Marines arrived at Tehran airport at about 2000. Marines from the Embassy were there to meet us, but had to leave to beat curfew. The seven of us made it through customs just before 2100, and were instructed to take special taxis which were allowed to travel the streets after curfew. It really hadn't dawned on us what we were getting ourselves into by traveling the streets during Martial Law. We were stopped twice on the way to the Embassy, and each time had H&K G3 automatic rifles stuck in our chests and had to show our American passports before we were allowed to proceed.

A conflict was taking place in Iran between the Shah's elected Prime Minister Bahktiar and exiled religious leader Ayatollah Khomeini. The revolution began the day after the army attacked the Doshan Tappen Air Force base at Farahabad on the morning of 10 February 1979. Iranian Air force cadets were staging anti and pro-Khomeini demonstrations when the Iranian Army was called in to restore order. Several people were rumored killed.

The next morning thousands of Iranians rushed out to the Air Force base, stormed the gates, and seized thousands of automatic weapons and grenades. It was a terrifying sight to see trucks and cars loaded with people carrying guns driving down the street past the United States Embassy. They were all yelling "Yankee go home."

The Marines at the Embassy handled the situation with a professional attitude as our mission was clearly defined as "to protect American interests and classified material abroad." This was the first time in my career in the Marine Corps where I had been subjected to this negative hostility towards Americans. It was an education in itself, and I learned a lot about myself. I learned to handle stress. I was able to define my responsibility as an NCO, as I was a Sergeant at the time and had about ten Marines under my direction. I learned as much about the security of the Embassy as possible, including the location of all the fire and intruder alarms, studying emergency and evacuation plans, and offering suggestions on how to improve these plans and improve security at the Embassy.

Embassy Marines being taken prisoner by Iranian militants in Teheran.

When I arrived in Tehran we had only thirteen watchstanders (it would eventually increase to twenty) and the Embassy had the additional protection of about 150-200 Iranian soldiers and security police. In addition to manning our normal posts we patrolled the Embassy grounds to check the gates and walls - as well as ensuring the Iranian soldiers were awake on their posts as they had a tendency to fall asleep.

When the Shah of Iran left the country and Khomeini arrived from Paris, relations between the Khomeini dictatorship and the American government were strained even more as anti-Americanism increased to the point where the Iranian army was forced to leave the American Embassy grounds or face severe punishment from higher military authority. With their departure, security of the Embassy and grounds was left to the Marine Detachment. Posts were set up around the perimeter, on top of the walls, and at the different gates as strictly surveillance posts to watch the streets. It was not unusual to stand post on the perimeter for twenty-four hours without a break.

On the morning of February 14, 1979 I was posted at the main gate along with two other Marines, Corporals Motten and Avery, and as I stepped up to the gate I noticed a police officer directing traffic away from the Embassy. About that time we heard gunfire, and radio reports of armed men on top of the walls coming into the complex. Permission was asked five times to throw gas.

Permission was finally granted by Ambassador Sullivan, but we realized gas wouldn't help against armed attackers. Since we hadn't encountered any of the attackers at our post we threw tear gas over the wall into the street where we could hear a lot of people yelling and running about. Then something happened which really blew my mind, and made me realize we weren't dealing with the army - but amateurs. Somebody threw a Molotov cocktail at me, and it broke at my boots. I looked down and saw the broken glass, the liquid, and the cloth. The idiot had forgotten to light it!

After we had exhausted our supply of tear gas canisters it

was time to move out and attempt to get back inside the Chancery. We gathered up #9 skeetshot ammunition for our shotguns, and proceeded deeper inside the compound around to the north side of the Chancery to a parking area for the staff where we took cover. At this time we experienced our first sniper fire coming from an eight story school located across from the motor pool area on the southwest corner of the compound. The attack had been underway for about twenty minutes, and the Marines were requesting permission to fire - only to realize it was a hopeless request because our shotguns, with #9 skeet, wouldn't do us any good against snipers. What I would have given for an M16!

Ambassador Sullivan finally gave us permission to fire if our lives were in danger. Well, our lives *were* in danger, but we really hadn't seen the enemy until we got around to the entrance of the Chancery. And the enemy we did see were too far away.

Once we got around the parked cars we discovered a couple of embassy employees hiding behind the vehicles and we helped them inside the Chancery. We did get off a couple rounds of cover fire as we were knocking on the door to get in. Once inside we took cover along with about thirty other people. Radio transmissions revealed two Marines were trapped inside the Ambassador's residence, so two Marines and a security officer grabbed weapons and were about to exit out thru the east entrance when the door was bombarded with machine gun fire. The two inch thick door was riddled full of holes, and any thought of attempting to exit that way was forgotten.

When the attackers came over the wall it was the Marines' responsibility to keep them away from the Chancery in order to give Embassy employees more time to destroy classified material. Our emergency and evacuation plan was that once we were inside the Chancery we were to make our way up to the top floor and gut the three lower floors with tear gas. Once upstairs we made our way to the communications vault, where people were still destroying classified material. Within an hour the attackers entered the building and walked straight into the gas. When they made it up to the third floor, they all had tears in their eyes from sucking tear gas. Ambassador Sullivan ordered the Marines to drop our weapons inside an office adjacent to the Ambassadors suite and give up without a fight. That decision probably saved a lot of lives, including the Marines.

After we had been frisked by the Iranian attackers we were led downstairs through all the teargas we had dispensed earlier to the basement, and then outside to the motor pool area. Once outside we were lined up against a wall and met by thirty photographers and reporters all taking photos and running around. The most frightening sight was that directly behind the photographers was a line of gunmen all armed with rifles and AK-47s in a firing squad formation. My thoughts at that time were they were going to move the press to the side and aim in. Thankfully, a spokesman for Khomeini arrived and ordered the attackers to leave the area. So, with weapons in hand, the attackers boarded buses and departed the U.S. Embassy compound and probably went on to another building in the city and shot it up as well.

Meanwhile we were all escorted to the Ambassador's residence for a couple of hours, and then allowed to move over to the embassy restaurant to eat and relax and comfort each other. It was pretty wild.

After the takeover of the Embassy, and up to the time the Marines were actually evacuated, our responsibilities included assisting Americans preparing to leave the country. Pan American Airlines volunteered to fly 747s into Tehran and help evacuate the over seven thousand Americans still there, so we had to assist with that as well as keep an eye on captives who enjoyed going around shooting up the Embassy or ransacking the Americans living quarters.

All of the original twenty Marines were evacuated and sent back to the States for debriefing, and were eventually sent out to another post. I chose to move on to Geneva Switzerland where I served for two years. Later on in 1979, as relations between the United States and Iran improved, thirteen other Marines were sent back to Tehran to stand post at the gates of the U.S. Embassy. Those same Marines, along with about thirty-nine other Americans, were taken hostage once again in November of 1979 and held captive for over fourteen months before they were released. That was about fourteen months too long, in my opinion, as we had an administration which didn't have the guts to send the Marines or another elite force over there to get them back. Instead they sent an ill-conceived joint military task force over to the Iranian desert, and that was where it ended - in the desert.

THAT WHICH DOES NOT
Kill You

MSG Detachment Tehran, Iran

Tim Mitchell

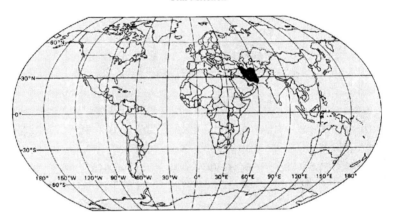

I have long been an advocate of giving SERE training – that is Survival, Evasion, Resistance and Escape instruction – to Marine Security Guards. I realize it would be impossible to send all MSGs to the full school, but it might be possible to add an abbreviated "in-house" version to the end of MSG School, or perhaps only send those being assigned to high risk posts. After all, they are in effect operating "behind enemy lines."

Paul Lewis says the frequent pain he feels in his wrists and shoulders gives him constant reminders of his ordeal as a hostage at the U.S. Embassy in Iran twenty-five years ago.

As I get older, what I thought were minor injuries don't

seem so minor anymore," Lewis said. "I need to have two or three surgeries due to the injuries I suffered from handcuffs on my wrists, and bindings on my shoulders."

Lewis, a former homecoming king and football star at Homer High School, was working as a Marine embassy guard when he became one of sixty-six Americans who were held in captivity for 444 days.

In February of 1979, the exiled Ayatollah Khomeini returned to Tehran and rose to power on a wave of popular discontent. Ten months later militants overran the embassy and took Lewis and other Americans captive while demanding the extradition of the Shah, who was in the United States for cancer treatment.

"The experience probably made me a stronger person. It gave me some strength of character that most people don't have, and it allowed me to overcome adversity," said Lewis, who is the head of his own financial advisory company in Urbana.

One thing Lewis doesn't want to forget from his time as a hostage is the friendships he made with his fellow hostages. He said he remains friends with many of them to this day, and they frequently send e-mails to one another - but Lewis said he would like to forget about the circumstances that led to his capture.

"Our embassy had been giving warnings to the Department of State about the unrest in Iran, but those warnings were ignored," he said.

Lewis said he believes he was able to survive the ordeal by focusing on surviving one day at a time.

"When you are in a situation like that, you have to focus only on the moment," Lewis said. "There was nothing I could do about the past. I could only deal with the present."

Lewis said he is very critical of current policy regarding

Iran. "We aren't doing enough to stop them. We've made it virtually no cost for Iranians to participate in terrorist activities. For example, the head terrorist who held us hostage is now the Iranian Ambassador to Syria."

But Lewis said he totally supports the war in nearby Iraq. "Iraq was run by a terrorist organization, and we thought there was the possibility of imminent danger," he said. "I am not opposed to actions against terrorist organizations. If we had chosen to ignore the situation, Iraq would have grown stronger, and the world would become more dangerous for all of us."

Lewis said he wishes the United States would have learned lessons from the takeover of the American embassy in Iran. "Terrorism has grown larger since 1979. Terrorist groups know there is a low cost to participate in these activities. We need to make the cost of terrorism high and immediate, whether the incidents are big or small."

Lewis said he also supports the U.S. policy against al-Qaida. "We know these people have killed Americans. What we are doing is morally justified."

He and the other hostages have filed suit against their captors under congressional legislation passed in 1996 which holds host countries responsible for terrorist activities. "The frustrating thing is that our own Department of State has intervened in federal court to stop our action," Lewis said. "We just want to hold the people who wronged us accountable for their dastardly actions."

WE'RE GOING TO DIE HERE

MSG Detachment Islamabad, Pakistan

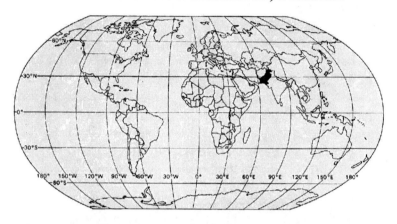

When I arrived in Brazzaville it was considered a "quiet" post much like Islamabad, but it too erupted in violence shortly after my arrival - necessitating the evacuation of all dependents and non-essential personnel. This article goes into a lot of detail about the causes of the Islamabad incident, and should give the reader a great deal of insight with regard to Islamo-fascist terrorism.

It was a small riot in a year of upheavals, a passing thunderclap disgorged by racing skies.

When the mob broke in William Putscher, a thirty-two-year-old American government auditor, was eating a hot dog. He had decided to lunch in the club by the swimming pool of the serene thirty-two-acre United States embassy compound in Islamabad, Pakistan. The embassy employed about 150 diplomats, spies, aid workers, communications specialists, assorted administrators, and a handful of U.S. Marines.

"Carter dog!" the rioters shouted, referring to American President Jimmy Carter. "Kill the Americans!" Putscher abandoned his meal and hid in a small office until the choking fumes of smoke and gasoline drove him out, and a raging protestor threw a brick in his face as he emerged. Another hit him on the back of his head with a pipe. They stole two rings and his wallet, hustled him into a vehicle, and took him three miles away to concrete dormitories at Quaid-I-Azam University. There student leaders of Pakistan's elite graduate school, fired by visions of a truer Islamic society, announced that Putscher would be tried for crimes "against the Islamic movement." It seemed to Putscher that he "was accused of just being an American."

It was November 21, 1979. As the riot erupted in Pakistan, forty-nine Americans sat imprisoned in the United States embassy in Tehran, trapped by Islamic radical students and Iranian revolutionary militia who announced that day a plan to murder the hostages by suicide explosions if any attempt was made to rescue them. In Mecca, Saudi Arabia - the holiest city in the Islamic world - Saudi national guardsmen encircled the Grand Mosque in pursuit of a failed theology student who had announced that he was the Mahdi, or Savior, dispatched to Earth by Allah as forecast in the Koran. To demonstrate their faith, the aspiring Mahdi's followers had opened fire on worshipers with automatic weapons. Just outside Washington, President Jimmy Carter prepared for Thanksgiving at Camp David. By day's end, he would have endured the first death by hostile fire of an American soldier during his presidency.

Inside the CIA station on the clean and carpeted third floor of the Islamabad embassy, the deputy chief of station, Bob Lessard, and a young case officer, Gary Schroen, checked the station's incinerator and prepared to burn classified documents. For situations like this, in addition to shredders, the station was equipped with a small gas-fed incinerator with its own chimney. Lessard sorted through case files and other classified materials, preparing if necessary to begin a burn.

Lessard and Schroen were both Persian-speaking veterans of service in Iran during the 1970s. Schroen, who had grown up in East St. Louis as the son of a union electrician, was the first member of his family to attend college. He had enlisted in the army in 1959, and was discharged honorably as a private. "I have a problem with authority," he told friends by way of explanation of his final rank. He kicked around odd jobs before joining the CIA in 1969, an agency full of people

171

who had problems with authority. As deputy chief of station, Bob Lessard was Schroen's boss, but they dealt with each other as colleagues. Lessard was a tall, athletic, handsome man with thinning hair and long sideburns. He had arrived at the Islamabad station feeling as if his career was in the doghouse. He had been transferred from Kabul, where an operation to recruit a Soviet agent had gone sour. An intermediary in the operation had been turned into a double agent without Lessard's knowledge, and the recruitment had been blown. Lessard had been forced to leave Afghanistan, and while the busted operation hadn't been his fault, he had landed in Islamabad believing he needed to redeem himself.

Life undercover forced CIA case officers into friendships with one another. These were the only safe relationships - bound by membership in a private society, unencumbered by the constant need for secrecy. When officers spoke the same foreign languages and served in the same area divisions, as Lessard and Schroen did, they were brought into extraordinarily close contact. To stay fit, Lessard and Schroen ran together through the barren chaparral of the hills and canyons around Islamabad. In the embassy they worked in the same office suite. Watching television and reading classified cables, they had monitored with amazement and dismay the takeover of the American embassy in Iran a few weeks earlier. Together they had tracked rumors of a similar impending attack on the U.S. Embassy in Islamabad. That Wednesday morning they had driven together into the Pakistani capital to check for gathering crowds, and they had seen nothing to alarm them.

Now, suddenly, young Pakistani rioters began to pour across the embassy's walls.

The Islamabad CIA station chief, John Reagan, had gone home for lunch, as had the American Ambassador to

Pakistan, Arthur Hummel. They missed the action inside the embassy that afternoon, but soon began to rally support from a command post at the British embassy next door.

Looking out windows, Schroen and Lessard could see buses pulling up before the main gate. Hundreds of rioters streamed out and jumped over sections of the embassy's perimeter protected by metal bars. One gang threw ropes over the bars and began to pull down the entire wall.

A group of hardcore student protestors carrying Lee Enfield rifles and a few pistols appeared on the lawns fronting the embassy's redbrick facade. One rioter tried to imitate Hollywood films by shooting an embassy gate lock with a pistol. As the American side later reconstructed events, the bullet ricocheted and struck protestors in the crowd. The rioters now believed they were being fired upon by U.S. Marines posted on the roof, and they began to shoot. Under their rules of engagement, the six Marine guards at the embassy that day could only fire their weapons to save lives. They were overwhelmed quickly, and outnumbered massively.

The Marines had always considered Islamabad a quiet posting. From the embassy's roof they could watch cows grazing in nearby fields. Master Gunnery Sergeant Lloyd Miller, a powerfully built Vietnam veteran who was the only member of his family to leave his small hometown in California, had seen nothing since his arrival in Pakistan a year earlier that even remotely compared to the battlefields around Danang. In July there had been a protest, but it wasn't much of one: "They sang a few songs, and chucked a few rocks. Then they went away." To pass the time, Miller and the Marines under his command drilled regularly. They practiced keeping modest-sized crowds out of the embassy compound, and even rehearsed what would happen if one or

173

two intruders found their way inside the building - but they had no way of preparing for what they now faced - wave upon wave of armed rioters charging directly toward their post in the lobby. Miller could see bus after bus pulling up near what was left of the front gates, but with only two security cameras on the grounds he could not assess just how pervasive the riot had become. He sent two of his Marines to the roof to find out.

Inside the embassy hallways only minutes later, shouts went up: "They shot a Marine!" In the CIA station Lessard and Schroen grabbed a medical kit and ran up the back stairway near the embassy's communications section. On the roof a cluster of embassy personnel knelt over the prone six-foot-six-inch figure of blond twenty-year-old Corporal Stephen Crowley of Port Jefferson Station, which is on Long Island in New York. He was a chess enthusiast and cross-country runner who had enlisted in the Marines two years before. Miller organized a makeshift stretcher from a slab of plywood lying close by. Crouched down low to avoid the bullets that whizzed overhead, they lifted Crowley onto the plywood and scampered toward the stairs. The CIA men held Crowley's head. The wound was life-threatening, but he might still be saved if they could get him out of the embassy and to a hospital. The stretcher bearers reached the third floor and headed toward the embassy's secure communications vault where the State Department and the CIA each had adjoining secure code rooms to send cables and messages to Washington and Langley. Emergency procedures dictated that in a case like this embassy personnel should lock themselves behind the communications vault's steel-reinforced doors to wait for Pakistani police or army troops to clear the grounds of attackers. It was now around one o'clock in the afternoon, and the riot had been raging for nearly an hour. Surely Pakistani reinforcements would not be long in coming.

Quaid-I-Azam University's campus lay in a shaded vale about three miles from the American embassy. A four-cornered arch at the entrance pointed to a bucolic expanse of low-slung hostels, classrooms, and small mosques along University Road. A planned, isolated, prosperous city laid out on geometrical grids, Islamabad radiated none of Pakistan's exuberant chaos. A Greek architect and Pakistani commissioners had combined to design the capital during the 1960s, inflicting a vision of shiny white modernity on a government hungry for recognition as a rising nation. Within Islamabad's antiseptic isolation, Quaid-I-Azam University was more isolated still. It had been named after the affectionate title bestowed on Pakistan's founding father, Mohammed Ali Jinnah, the "Father of the Nation." Its students plied walkways shaded by weeping trees beneath the dry, picturesque Margalla Hills, several miles from Islamabad's few shops and restaurants. During much of the 1970s the university's culture had been Western in many of its leanings. Women could be seen in blue jeans, men in the latest sunglasses and leather jackets. This partly reflected Pakistan's seeming comfort in an era of growing international crosscurrents. Partly, too, it reflected the open, decorative cultural styles of Pakistan's dominant ethnic Punjabis. In Lahore and Rawalpindi, hotels and offices festooned in electric lights winked at passersby. Weddings rocked wildly through the night with music and dance. While the ethnic mix was different, in coastal Karachi social mores were perhaps even more secular, especially among the country's business elites. For the most part, Quaid-I-Azam's students expressed the fashion-conscious edges of this loose, slightly licentious stew of Islamic tradition and

175

subcontinental flair.

More recently however, an Islamist counterforce had begun to rise at the university. By late 1979 the student wing of a conservative Islamic political party, Jamaat-e-Islami (the Islamic Group, or alternatively, the Islamic Society) had taken control of Quaid-I-Azam's student union. The Jamaat student activists, while a minority, intimidated secular-minded professors and students, and shamed women who adopted Western styles or declined to wear the veil. Like their elder political leaders, Jamaat students campaigned for a moral transformation of Pakistani society through the application of Islamic law. Their announced aim was a pure Islamic government in Pakistan. The party had been founded in 1941 by the prominent Islamic radical writer Maulana Abu Ala Maududi, who advocated a Leninist revolutionary approach to Islamic politics, and whose first book, published in the late 1920s, was titled *Jihad in Islam*. Despite its leaders' calls to arms, Jamaat had mainly languished on the fringes of Pakistani politics and society, unable to attract many votes when elections were held and unable to command much influence during periods of military rule. Maududi had died just weeks earlier, in September of 1979, with his dream of an Islamic state in Pakistan unrealized. Yet at the hour of his passing his influence had reached a new peak, and his followers were on the march. The causes were both international and local.

Because it had long cultivated ties to informal Islamic networks in the Persian Gulf and elsewhere, Jamaat-e-Islami found itself afloat during the 1970s on a swelling tide of what the French scholar Gilles Kepel would later term "petro dollar Islam," a vast infusion of proselytizing wealth from Saudi Arabia arising from the 1973 oil boycott staged by the Organization of Petroleum Exporting Countries (OPEC). The boycott sent global oil prices soaring. As angry Americans pumped their Chevrolets with dollar-a-gallon gasoline which was a lot back then) they filled Saudi and other Persian Gulf treasuries with sudden and unimagined riches. Saudi Arabia's government consisted of an uneasy alliance between its royal family and its conservative, semi-independent religious clergy. The Saudi clergy followed an unusual, puritanical doctrine of Islam often referred to as "Wahhabism," after its founder, Mohammed ibn Abdul Wahhab, an eighteenth-century desert preacher who regarded all forms of adornment and modernity as blasphemous. Wahhabism's insistent severity stood in opposition to many of the artistic and cultural traditions of past Islamic civilizations, but it was a determined faith - and now overnight, an extraordinarily wealthy one. Saudi charities and proselytizing organizations such as the Jedda-based Muslim World League began printing Korans by the millions as the oil money gushed. They endowed mosque construction across the world and forged connections with like-minded conservative Islamic groups from Southeast Asia to the Maghreb, distributing Wahhabi-oriented Islamic texts and sponsoring education in their creed.

In Pakistan, Jamaat-e-Islami proved a natural and enthusiastic ally for the Wahhabis. Maududi's writings, while more anti-establishment than Saudi Arabia's self-protecting monarchy might tolerate at home, nonetheless promoted many of the Islamic moral and social transformations sought by Saudi clergy.

By the end of the 1970s Islamic parties like Jamaat had begun to assert themselves across the Muslim world as the corrupt, failing reigns of leftist Arab nationalists led youthful populations to seek a new cleansing politics. Clandestine, informal, transnational religious networks such as the

Muslim Brotherhood reinforced the gathering strength of old-line religious parties such as Jamaat. This was especially true on university campuses, where radical Islamic student wings competed for influence from Cairo to Amman to Kuala Lumpur. When Ayatollah Khomeini returned to Iran and forced the American-backed monarch Shah Mohammed Reza Pahlavi to flee early in 1979, his fire-breathing triumph jolted these parties and their youth wings, igniting campuses in fevered agitation. Khomeini's minority Shiite creed was anathema to many conservative Sunni Islamists, especially those in Saudi Arabia, but his audacious achievements inspired Muslims everywhere.

On November 5, 1979 Iranian students stormed the U.S. embassy in Tehran, sacked its offices, and captured hostages. The next morning in Islamabad's serene diplomatic quarter near the university, local Iranians draped their embassy with provocative banners denouncing the United States and calling for a global Islamic revolution against the superpowers. The student leaders of Jamaat were enthusiastic volunteers. Although the party's older leaders had always focused their wrath on India - motivated by memories of the religious violence that accompanied Pakistan's birth - the new generation had its sights on a more distant target: the United States. Secular leftist students on campus also denounced America. Kicking the American big dog was an easy way to unite Islamist believers and nonbelievers alike.

Jamaat's student union leaders enjoyed an additional pedigree - they had lately emerged as favored political protégés of Pakistan's new military dictator, General Mohammed Zia-ul-Haq. The general had seized power in July 1977 from the socialist politician Zulfikar Ali Bhutto, father of future prime minister Benazir Bhutto. Despite personal appeals for clemency from President Carter and many other world leaders, Zia sent Zulfikar Ali Bhutto to the gallows in April of 1979. Around the same time American intelligence analysts announced that Pakistan had undertaken a secret program to acquire nuclear weapons. Zia canceled elections and tried to quell domestic dissent. Shunned abroad and shaky at home, he began to preach political religion fervently, strengthening Jamaat in an effort to develop a grassroots political base in Pakistan. In the years to come, engorged by funds from Saudi Arabia and other Gulf emirates, Jamaat would become a vanguard of Pakistan's official and clandestine Islamist agendas in Afghanistan, and later Kashmir.

On October 21, 1979 Zia announced that he intended to establish "a genuine Islamic order" in Pakistan. Earlier in the year he had approved Islamic punishments such as amputations for thieves and floggings for adulterers. These turned out to be largely symbolic announcements, since the punishments were hardly ever implemented. Still, they signaled a new and forceful direction for Pakistan's politics. Conveniently, since he had just aborted national polls, Zia noted that "in Islam there is no provision for Western-type elections." Jamaat's leaders defended him, and its student wing, an eye cocked at the celebrated violence of Iranian student radicals, prepared to demonstrate its potency.

In this incendiary season arrived a parade of apparent mourners wearing red handbands and shouldering coffins at Mecca's holy Grand Mosque in the western deserts of Saudi Arabia. The picture they presented to fellow worshipers at dawn on Tuesday, November 20 was not an uncommon one because the mosque was a popular place to bless the dead. There would soon be more to bless. The mourners set their coffins down, opened the lids, and unpacked an arsenal of assault rifles and grenades.

Their conspiracy was born from an Islamic study group at Saudi Arabia's University of Medina during the early 1970s. The group's leader, Juhayman al-Utaybi, had been discharged from the Saudi national guard. He persuaded several hundred followers - many of them Yemenis and Egyptians who had been living in Saudi Arabia for years - that his Saudi brother-in-law, Mohammed Abdullah al-Qahtani, who had once studied theology, was the Savior returned to Earth to save all Muslims from their depredations. Juhayman attacked the Saudi royal family. Oil-addled royal princes had "seized land" and "squandered the state's money," he proclaimed. Some princes were "drunkards" who "led a dissolute life in luxurious palaces." He had his facts right, but his prescriptions were extreme. The purpose of the Mahdi's return to Earth was "the purification of Islam" and the liberation of Saudi Arabia from the royal family. Signaling a pattern of future Saudi dissent, Juhayman was more puritan than even Saudi Arabia's officially sanctioned puritans as he sought bans on radio, television and soccer. That November morning, impatient with traditional proselytizing, he chained shut the gates to the Grand Mosque, locking tens of thousands of stunned worshipers inside. The mosque's imam declined to ratify the new savior. Juhayman and his gang began shooting, and dozens of innocent pilgrims fell dead.

Saudi Arabia did little in the early hours of this bizarre uprising to clarify for the Islamic world who was behind the assault. Every devout Muslim worldwide faced Mecca's black, cube-shaped Kaaba five times a day to pray. Now it had been captured by usurping invaders. But who were they, and what did they want? Saudi Arabia's government was disinclined to publicize its crises. Saudi officials were themselves uncertain initially about who had sponsored the attack. Fragmented eyewitness accounts and galloping rumors leaped from country to country, continent to continent. In Washington, Secretary of State Cyrus Vance dispatched an overnight cable to U.S. embassies worldwide on that Tuesday night, urging them to take precautions as the Mecca crisis unfolded. The State Department had painfully learned only weeks earlier about the vulnerability of its compounds and the speed at which American diplomats could face mobs inflamed by grievances, real and imagined.

Ambassador Hummel in Islamabad sorted through these cabled cautions the next morning. He did not regard Islamic radicalism as a significant threat to Americans in Pakistan. It never had been before. Still, the Islamabad CIA station had weeks earlier picked up indications from its sources that students at Quaid-I-Azam might be planning demonstrations at the embassy in support of the Iranian hostage takers in Tehran. As a result, Hummel had requested and received a small contingent of about two dozen armed Pakistani police, over and above the embassy's normal security force.

That squad was in place on Wednesday morning when rumors began to circulate in Islamabad, and later on local radio stations, that the United States and Israel stood behind the attack at the Grand Mosque. The rumor held that Washington and Tel Aviv had decided to seize a citadel of Islamic faith in order to neutralize the Muslim world. Absurd on its face, the rumor was nonetheless received as utterly plausible by thousands if not millions of Pakistanis. The Voice of America reported that as the riot in Mecca raged, President Carter had ordered U.S. Navy ships to the Indian Ocean as a show of force against the hostage takers in Tehran. With a little imagination it wasn't hard to link the two news items. As the students at Quaid-I-Azam made their protest plans *The Muslim*, an Islamabad daily, published a special edition that referred to the "two hostile actions against the Muslim world... by the Imperialists and their stooges."

General Zia had plans that day to promote civic advancement through Islamic values. He had decided to spend most of the afternoon in teeming Rawalpindi, adjacent to Islamabad, riding about on a bicycle. Zia intended to hand out Islamic pamphlets and advertise by example the simple virtues of self-propelled transport. And of course, where the military dictator went, so went most of Pakistan's military and security establishment. When the first distress calls went out from the U.S. embassy later that day, much of Pakistan's army brass was unavailable. They were pedaling behind the boss on their bicycles.

Gary Schroen stood by the window of his office preparing to close the curtains when a Pakistani rioter below raised a shotgun at him and blasted out the plate glass. He and the young Marine beside him had spotted the shooter just early enough to leap like movie stuntmen beyond the line of fire. The shotgun pellets smashed into the CIA station's plaster walls. They had no time now to destroy classified documents. Schroen and Lessard locked their case files and disguised materials in the station suite behind a vault door, grabbed a pair of pump-action Winchester 1200 shotguns from a Marine gun case, and headed to the third-floor code room vault.

By about 2 PM, 139 embassy personnel and Pakistani employees had herded themselves inside, hoping for shelter from the mob. Within the vault a young political officer had cleared off a desk and was busy writing by hand the FLASH cable that would announce the attack to Washington. As he wrote, embassy communications officers destroyed cryptography packages one by one to prevent them from falling into the hands of rioters. The vault echoed with the sound of a sledgehammer rhythmically descending on CIA code equipment.

The wounded Marine, Stephen Crowley, lay unconscious and bleeding on the floor, tended by an embassy nurse. He was breathing with help from an oxygen tank. Crowley had been shot in the riot's early moments, and by now the protestors had swollen in number and anger and had begun to rampage through every corner of the compound. They hurled Molotov cocktails into the chancery's lower offices, setting files and furniture on fire. Entire wings of the building leaped in flames, particularly the paper-laden budget and finance section located directly underneath the communications vault, which began to cook like a pot on a bonfire. Onlookers at the British embassy estimated that at the height of the action, fifteen thousand Pakistani rioters swarmed the grounds.

Marine Master Gunnery Sergeant Miller - or the "Gunny," as he was called - directed the defense from his post in the lobby. There he watched as rioters rushed through the now mangled front door no more than fifteen feet away. They scurried into the lobby carrying bundles of wood, buckets of gasoline, and matches. Miller repeatedly requested permission for his men to fire on the arsonists, but each time the embassy's administrative counselor, David Fields, denied the request on the grounds that shooting would only further incite the riot. Miller had to content himself with rolling out more tear gas canisters as fire engulfed the building he was sworn to protect.

When the lobby had completely filled with smoke, the Marines retreated upstairs to join the rest of the embassy staff in the third-floor vault. Just before going in, they dropped a few final tear gas canisters down each of the stairwells in the hope that would dissuade the rioters from climbing to the embassy's last remaining refuge.

Outside at the motor pool the rioters poured gasoline into embassy cars and set them burning one after another. In all, more than sixty embassy vehicles would go up in flames. Some rioters attacked the embassy residences, a cluster of modest brick town houses that were home to midlevel American personnel and their families. Quaid-I-Azam University student leaders rounded up a group of hostages from these quarters and announced their intention to drive them to the campus to put them on trial as American spies. An enterprising Pakistani police lieutenant - one of the few guards who had refused to surrender his weapon to the mob in the riot's earliest moments - pretended to go along with the students' plan, loaded the hostages into a truck, and promptly drove them off to safety. He was not the only Pakistani to risk himself for the Americans. At the American School in Islamabad several miles away from the embassy, a retired army colonel armed an impromptu squad of Pakistani guards with cricket bats and broomsticks. They successfully beat off rioters who attacked the school while children lay cowering in locked rooms. Although these and other individuals acted heroically, Pakistan's government did not. Despite dozens of pleas from Ambassador Arthur Hummel and CIA station chief John Reagan, hour after hour passed and still no Pakistani troops or police arrived to clear the rioters. By mid-afternoon enormous black clouds of gasoline-scented smoke poured out from the American compound, visible from miles away.

Many of the rioters joined the melee spontaneously, but as the rampage unfolded it also revealed evidence of substantial coordinated planning. On the embassy grounds CIA personnel spotted what appeared to be riot organizers wearing distinctive sweater vests and carrying weapons. Some were Arabs, likely members of the sizable Palestinian population at Quaid-I-Azam. The speed with which so many rioters descended on the embassy also suggested advanced preparation. Thousands arrived in government-owned Punjab Transport Corporation buses. Rioters turned up nearly at once at multiple American locations - the embassy compound, the American School, American information centers in Rawalpindi and Lahore, and several American businesses in Islamabad. Professors at Quaid-I-Azam later reported that some students had burst into classrooms very early in the morning, before the rumor about American involvement in the Grand Mosque uprising had spread very far, shouting that students should attack the embassy to take vengeance in the name of Islam.

Around 4 PM Pakistani army headquarters finally dispatched a helicopter to survey the scene. It flew directly above the embassy, its whirring rotors fanning flames that raked the building. Then the helicopter flew away. Zia's spokesmen later said the smoke had been too thick to make a visual assessment. The CIA reported that its sources in Zia's circle told a different story. When the helicopter returned to base, the crew advised Zia that the fire in the embassy was so hot and so pervasive that there was no way the American personnel inside could have survived. Since it seemed certain that the Americans had all been killed, there was no sense in risking further bloodshed - and a possible domestic political cataclysm - by sending army troops to forcibly confront the Islamist rioters. According to the CIA's later reports, Zia decided that since he couldn't save the Americans inside the embassy anyway, he might as well just let the riot burn itself out.

By this time the Americans and Pakistanis in the vault were nearing the end of their tolerance. They had been inside for more than two hours, and there was no rescue in sight. In the State Department's chamber they lay drenched with sweat and breathing shallowly through wet paper towels. Tear gas had blown back to the third floor, and some were gagging and vomiting. Temperatures rose as fires in the offices below burned hotter. Carpet seams burst from the heat. Floor tiles blistered and warped.

In the adjacent CIA code room, Miller, Schroen, Lessard, and a crew of CIA officers and Marine guards stared at a bolted hatch in the ceiling that led up to the roof. They wondered if they should try to force the hatch open and lead everyone to the fresh air above. A previous Islamabad station chief had installed the hatch for just this purpose, but about an hour into the attack the rioters had discovered the passageway. They pounded relentlessly on the iron lid with pieces of a brick wall they had torn apart, hoping to break in. Some rioters poked their rifles into nearby ventilation shafts and shot. The sound of bullets crashing down from above was occasionally punctuated by even more jolting explosions as the fire crept up on oxygen tanks stored elsewhere in the building.

The group in the code room listened to the metallic clanging on the hatch for about an hour. Then one of the CIA communications specialists, an engineer of sorts, came up with a plan to wire a heavy-duty extension cord into the iron cover. "Those guys up there, I'm going to electrocute them!" he announced gleefully, as one of his colleagues later recalled it. He stripped to the waist and began to sweat as he attached large alligator clips to the hatch. "Now I'm going to plug this baby in, and the electricity's going to kill them." Filthy and covered with bits of shredded documents, he thrust the plug into the wall. Four hundred volts of current seemed to fly up to the hatch, bounce off, and fly right back into the wall where it exploded in sparks and smoke. "Goddamn it! The resistance is too much!"

The idea had seemed dubious from the beginning - the device wasn't even grounded properly - and there was laughter for the first time all afternoon when it failed, but what other options did they have? The heat had grown unbearable inside the vault. "What are we going to do?" they asked. "They're up there. What are we going to do?"

Another hour passed. Slowly the hatch bent under the rioters' bricks. The concrete around it began to crumble into the code room. The CIA officers and Marines estimated they had about thirty minutes before the cover collapsed, but suddenly the banging stopped and the voices on the roof quieted. After a few minutes of silence the Gunny decided, "Let's open the hatch and we'll face what happens." The Ambassador had given them the go-ahead to fire first to maintain security in the vault, and they had enough weaponry to make it a battle if it came to that.

Lessard and Schroen climbed ladders and popped the hatch halfway off. Their colleagues crouched below, shotguns primed. There were half a dozen of them, and they were ready to shoot as soon as the rioters poured in.

"Guys, guys! When we open the hatch, if somebody's up there, we're going to drop down. Then shoot! Don't shoot first!" They worked out a plan for sequential firing.

Schroen looked across the ladder at Lessard. "We're going to die here if anybody..."

"Yeah, I think so, Gary."

But they couldn't open the hatch. They beat on the bolt, but the contraption was now so bent and warped that it wouldn't pop. They pushed and pushed, but there was nothing they could do.

The sun set on Islamabad, and the noises outside began to drift off into the chilly November air. It was now about 6:30 PM. Maybe the rioters were gone, or maybe they were lying in wait for the Americans to try to escape. David Fields, the administrative counselor, decided it was time to find out. He ordered the Gunny to lead an expedition out the third-floor hallway and up onto the roof. Fields told them they had the authority to fire on any rioters who got in their way.

Miller and his team of five sneaked out of the vault and into a hallway thick with smoke. They ran their hands along the curved hallway wall to keep track of their position and felt their way to the end where a staircase led to the roof. The locked metal door normally guarding access to the stairs had been torn off its hinges. The rioters had already been here.

With shotguns and revolvers locked and loaded, Miller cautiously guided his team up the stairs. As he poked his head out onto the roof, he fully expected a shoot-out. Instead, he saw a single Pakistani running toward him with hands raised high in the air and yelling, "Friend! Friend!" Miller gave the man a quick pat-down and found a copy of *Who's Who in the CIA* stuffed in one of his pockets, suggesting that student leaders had planned, Tehran-style, to arrest their own nest of spies. Miller took the book and told the straggler to get lost. The Gunny would not fire his weapon that day, nor would any of the Marines under his command. The riot had finally dissipated. During the last hour it had degenerated gradually into a smoky, sporadic carnival of looting.

A few minutes after the expedition party set out, those still inside the vault heard the sound of the hatch being wrenched from above. An enormous U.S. Marine with hands like mallets ripped it off its moorings. Soon everyone from the CIA code room was up on the roof and staring over the chancery walls. Through the halo of smoke that ringed the building they looked across the embassy grounds and saw bright leaping flames where some of their homes had once stood. All of the embassy compound's six buildings, constructed at a cost of twenty million, had been torched beyond repair.

Using bicycle racks stacked end to end, the Marines set up makeshift ladders and led the large group huddled in the vault to safety. It was now dark and cold, and the footing was precarious. Vehicle lights and embers from fires illuminated the ground in a soft glow. Some Pakistani army troops had finally arrived. They were standing around inside the compound, mostly watching.

When the last of those in the vault had been helped down, the Gunny turned to climb the ladder. The CIA men asked where he was going. "I've got to go get Steve," he said. "I'm not going to leave my man up there."

Minutes later he emerged with Crowley's inert form wrapped in a blanket, slung across his shoulder. Crowley had died when the oxygen supply in the vault ran out. In flickering light the Gunny carried the body down the ladder to the ground.

"All reports indicate all of the people in the compound have been removed and taken to safety thanks to the Pakistani troops," State Department spokesman Hodding Carter told reporters in Washington later that day. In a telephone call, President Carter thanked Zia for his assistance, and Zia expressed regret about the loss of life. The Pakistani Ambassador in Washington accepted the Americans' gratitude and noted that Pakistani army troops had reacted "promptly, with dispatch." Secretary of State Cyrus Vance hurriedly summoned ambassadors from thirty Islamic countries to discuss the Pakistan embassy attack and its context. Asked about the recent wave of Islamic militancy abroad, Vance said, "It's hard to say at this point whether a pattern is developing."

It took a day or two to sort out the dead and missing. Putscher, the kidnapped auditor, was released by the students at Quaid-I-Azam around midnight. They had called him "an imperialist pig" and found America guilty "of the trouble in Mecca and all the world's problems," but they decided in the end that he was personally innocent. He wandered back to the embassy, wounded and shaken.

Rescue workers found two Pakistani employees of the embassy in a first-floor office. They had died of apparent asphyxiation, and their bodies had been badly burned. In the compound's residential section workers found an American airman, twenty-nine-year-old Brian Ellis, lying dead on the floor of his fire-gutted apartment. A golf club lay beside him - he had apparently been beaten unconscious and left to burn.

On Friday a Pan American Airlines jumbo jet evacuated 309 nonessential personnel, dependents, and other Americans from Pakistan and back to the United States.

Saudi Arabian soldiers and French commandos routed the armed attackers at the Grand Mosque on Saturday in a bloody gun battle. The Saudis never provided an accounting of the final death toll, but most estimates placed it in the hundreds. Saudi interior minister Prince Naif downplayed the uprising's significance, calling the Saudi renegades "no more than a criminal deviation" who were "far from having any political essence." Surviving followers of the Mahdi, who had been shot dead, fled to the mosque's intricate network of basements and underground tunnels. They were flushed out by Saudi troops after a further week of fighting. The building contractor who had originally reconstructed the mosque for the Saudi royal family reportedly supplied blueprints that helped security forces in this final phase of the battle. The Bin Laden Brothers for Contracting and Industry were, after all, one of the kingdom's most loyal and prosperous private companies.

The American treasury secretary, William Miller, flew into the kingdom amid the turmoil. He hoped to reassure Saudi investors, who had about thirty billion on deposit in U.S. banks, that America would remain a faithful ally. He also urged the Saudi royal family to use their influence with OPEC to hold oil prices in check. Rising gasoline prices had stoked debilitating inflation and demoralized the American people.

Saudi princes feared the Mecca uprising reflected popular anxiety about small Westernizing trends which had been permitted in the kingdom during recent years. They soon banned women's hairdressing salons and dismissed female announcers from state television programs. New rules stopped Saudi girls from continuing their education abroad. Prince Turki al-Faisal, the Saudi intelligence chief, concluded the Mecca uprising was a protest against the conduct of all Saudis - the sheikhs, the government, and the people in general. There should be no future danger or conflict between social progress and traditional religious practices, Turki told visitors, as long as the Saudi royal family reduced corruption and created economic opportunities for the public.

In Tehran, the Ayatollah Khomeini said it was "a great joy for us to learn about the uprising in Pakistan against the USA. It is good news for our oppressed nation. Borders should not separate hearts." Khomeini theorized that "because of propaganda, people are afraid of superpowers, and they think that the superpowers cannot be touched." This, he predicted, would be proven false.

The riot had sketched a pattern that would recur for years. For reasons of his own, Pakistani dictator General Zia had sponsored and strengthened a radical Islamic partner - in this case Jamaat, and its student wing - that had a virulently anti-American outlook. This Islamist partner had veered out of control. By attacking the American embassy Jamaat had far exceeded Zia's brief, yet Zia felt he could not afford to repudiate his religious ally - and the Americans felt they could not afford to dwell on the issue. There were larger stakes in the U.S. relationship with Pakistan. In a crisis-laden, impoverished Islamic nation like Pakistan, which was on the verge of acquiring nuclear weapons, there always seemed to be larger strategic issues for the United States to worry about than the vague, seemingly manageable dangers of political religion.

On the night of the embassy's sacking, Zia gently chided the rioters in a nationally broadcast speech. "I understand that the anger and grief over this incident were quite natural," he said, referring to the uprising in Mecca, "but the way in which they were expressed is not in keeping with the lofty Islamic traditions of discipline and forbearance." As the years passed, Zia's partnership with Jamaat would only deepen.

The CIA and State Department personnel left behind in Islamabad felt deeply embittered. They and more than one hundred of their colleagues had been left to die in the embassy vault – after it had taken Pakistani troops more than five hours to make what was at maximum a thirty-minute drive from army headquarters in Rawalpindi. Had events taken a slight turn for the worse, the riot would have produced one of the most catastrophic losses of American life in U.S. diplomatic history.

The CIA's Islamabad station now lacked vehicles in

which to meet its agents. The cars had all been burned by mobs. Gary Schroen found an Quaid-I-Azam University jeep parked near the embassy, which had apparently been left behind by the rioters. Schroen hot-wired it so that he could continue to drive out at night for clandestine meetings with his reporting agents. Soon university officials turned up at the embassy to ask after the missing jeep - the university now wanted it back. Schroen decided that he couldn't afford to drive around Islamabad in a vehicle that was more or less reported as stolen, so he drove the jeep one night to a lake on Islamabad's outskirts, got out, and rolled it under the water. It was small satisfaction, but it was something.

THE FUN NEVER STOPS

MSG Detachment Monrovia, Liberia

Sergeant Caleb Eames

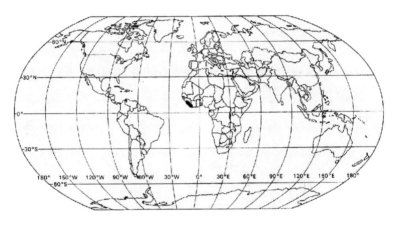

Liberia, officially the Republic of Liberia, is a country on the west coast of Africa, bordered by Sierra Leone, Guinea, and Côte d'Ivoire. Liberia, which means "Land of the Free," was founded as an independent nation by free-born and formerly enslaved African Americans. The founding of Liberia was privately sponsored by American religious and philanthropic groups, but the country enjoyed the support and unofficial cooperation of the United States government. Recently it has witnessed two civil wars, the Liberian Civil War (1989–1996), and the Second Liberian Civil War (1999–2003), which have displaced hundreds of thousands and destroyed its economy. On 12 April 1980 a successful military coup was staged by a group of noncommissioned Krahn officers led by Master Sergeant Samuel Kanyon Doe,

and they executed the President of nine years William R. Tolbert, Jr. in his mansion after torturing him for a period of several days. Constituting themselves as the People's Redemption Council, Doe and his associates seized control of the government and brought an end to Africa's "first republic." However in late 1989 a civil war began and on 5 August, 1990 Marines from a U.S. Task Force off the coast of Liberia began an evacuation code named Operation SHARP EDGE which eventually rescued 2,690 people, including 330 U.S. citizens, from the war-torn capital city of Monrovia. In September of 1990 Doe was finally ousted and killed by the forces of faction leader Yormie Johnson and members of the Gio tribe. Prominent warlord Charles Taylor was then elected President in 1997. Taylor's brutal regime targeted several leading opposition and political activists, and his autocratic and dysfunctional government led to a new rebellion in 1999 - which continued until he was finally forced to resign in 2003. More than 200,000 people are estimated to have been killed in the civil wars. The following incident took place on September 18-30, 1998.

It all started on Friday as I walked out of the Marine House on my way to the embassy when a loud "WHUMMPP" reverberated throughout the compound. It was an RPG round detonating. The local government forces had decided to eliminate the rebel opposition who lived primarily on Camp Johnson Road, and the whole day automatic weapons fire intermixed with RPG rounds detonating echoed throughout the city. We knew there would be a problem when stray rounds started impacting on some of the buildings on the embassy compound.

We received frantic calls from Americans throughout the city asking for advice. We told them to remain inside and stay safe, as the violence was not directed toward Americans. Throughout that night and into the morning thousands of people fleeing the fighting showed up at the gate to the embassy. By noon on Saturday about twenty-five thousand refugees had made their way inside a housing compound across the street from the embassy, all seeking protection from the fighting. By Saturday afternoon the rebel leader, Roosevelt Johnson, somehow made it out of the fighting and arrived at the front gate to the embassy asking for asylum. Some security personnel then went out into the street with the acting Ambassador to negotiate a peaceful handover of Mr. Johnson to the ECOMOG peacekeeping force here. During the negotiations the Special Security Unit, the "secret police" of the government militia, started arriving in force. These SSU troops' mission was to find and kill Mr. Johnson and any supporters.

The Embassy compound in Monrovia, Liberia as it looked in 1998.

The street in front of the embassy was filled with about thirty vehicles, ECOMOG MOWAGS (armored personnel carriers), police vehicles, armored sedans, and men armed to the teeth with AK-47s and other automatic weapons and RPGs. At that point Mr. Johnson and several men loyal to him were still outside the embassy hidden behind a partition, which shielded them from the view from the street.

With American Security personnel still on the street, approximately seven men in a tightly stacked formation, moved around the partition with weapons raised. The gunfire started with a few pops, and then the noise of automatic fire burst out. An RPG round was fired toward the embassy, but fortunately it traveled directly over the building and landed in the ocean on the other side.

Later another RPG round was located on the grounds, undetonated. I would estimate 150 rounds were fired in a thirty-second period. This fire was all directed at Mr. Johnson and the Americans surrounding him as he went into the gate, through it, and into the embassy itself. With a hail of bullets around them the Americans dove for the turnstile gate while returning fire as best they could, having to go through a literal "Fatal Funnel" one by one. Mr. Johnson and his followers also jumped through along with them.

With all the shooting, only two Americans were wounded. The Liberians have no concept of aiming - they just spray the area hoping to hit something. One American was hit in the hip, and the other in the arm. Mr. Johnson came through unharmed, but one of his sons was shot in the shoulder and arm, another man was shot in the neck, and a third was hit several times in the chest. Several of Johnson's followers were shot down in the street, and the American security guards killed at least two SSU troops. Several local guard force employees were wounded, though none severely.

With the Americans coming through the gate, there was no way to return fire from the embassy itself without hitting one of our own. Local guards dragged the wounded into the embassy building with Marines assisting. The man shot in the chest bled to death on the floor behind Post One, the two Americans were treated as best as possible in the embassy gym, and Mr. Johnson and his people were put into a stairwell under Marine guard and given hasty treatment for their gunshot wounds. At this point, we received reports of men jumping the perimeter along the side of the embassy. Things got tense as we wondered if they were armed SSU or just refugees.

Later we learned they were twenty additional Johnson supporters who had climbed over the embassy razor-wire fence seeking refuge, and in the process inflicted severe cuts on their arms and legs. All of the Liberians were given first aid, and later we had doctors come in and treat their wounds. Gunfire continued in the streets outside the embassy for several hours, until ECOMOG brought in some heavy caliber machineguns and started laying down fire to disperse the fighters.

The decision was made to evacuate all non-essential personnel that afternoon to Sierra Leone. We also advised all Americans to leave Liberia, since the situation had worsened in town. For the next week and a half, all embassy personnel were confined to the embassy building. Tensions were high as no one knew for sure what to expect. Those were some long nights with every rumor demanding an appropriate defensive response, until reinforcements arrived to supplement our security several days later. We kept Mr. Johnson in the lobby of the Embassy the entire time, with an armed escort to use the bathroom and take showers. When word arrived that a deal had been made to get him out of the country, all of us were relieved.

Under Marine Guard, and with AP and wire service reporters watching and filming, Mr. Johnson and his followers - some limping with serious wounds - loaded into two helicopters. Things didn't end there. With apprehension about possible rogue forces looking for revenge, we stayed in the embassy for several more days.

Civilians boarding a CH-53E Super Stallion during a NEO.

Finally, on the 30th of September, we allowed people to move around the compound. I called some friends who work for Tearfund - when the whole thing started they had gone to the 'Save the Children' compound. The night the shooting occurred here at the Embassy some SSU soldiers went to the compound where they were staying, broke in the door, and started making demands on the premise they were searching for rebels. They fired shots into the ceiling to make the people comply, and then proceeded to take money and valuables from the house. It was quite a tense situation for everyone involved, but everyone got through it unharmed - thank God.

That story could be applied in many variations throughout the city, as the SSU troops ravaged innocent civilians. Things here are pretty calm now, but there is an uneasy underlying feeling especially when speaking of Americans. I think the government still holds a kind of grudge for the incident, and we fear another increase in tensions. It is estimated that between three and four hundred people were killed in a kind of "ethnic cleansing" of the rebel faction composed primarily of the Khran tribe during the two days of fighting. The rest of the city has returned to normal, and people continue going about their daily business. The only exception is the other day SSU troops broke into the Barclay Training Center where some Khran military Generals were. They summarily executed all of them on the spot. Then rumors started circulating that they were coming to Camp Johnson Road again, this time to shoot ANYONE who was around, regardless of what they were doing.

People started fleeing the city about 2 AM by the thousands, but by morning life was back to normal. We are still operating under heightened security, but things seem to be relatively quiet. We hope for the best, but ready ourselves for the worst.

As Marines before us did, and as many will in the future, the Marines on duty during this incident displayed the highest caliber of outstanding professionalism in the face of danger. They upheld the tradition and honor of the United States Marine Corps, and performed their mission of defending the United States Embassy and United States citizens in Monrovia, Liberia.

PRELUDE TO WAR

MSG Det Kuwait City, Kuwait

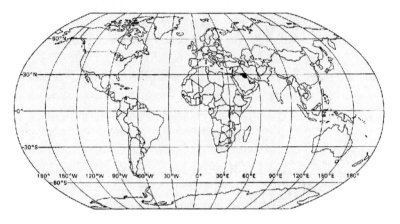

Some situations are a lot bigger than your op-plans, and when that happens about all you can do is batten down the hatches and wait for the cavalry to arrive. The Iraqis could have easily overwhelmed the embassy, but I suppose Saddam Hussein thought the U.S. would stay out of the Kuwaiti affair if he didn't antagonize us too much. He was wrong.

When the Iraqi Army invaded Kuwait in 1990 the United States ordered naval forces to the Persian Gulf, but they were making slow headway against heavy seas. In Kuwait City, on August 3, some of Saddam's Republican Guards were outside the U.S. Embassy and were threatening to go over the wall. The twenty or so people inside with Ambassador Howell were terrorized, but determined to fight with shotguns, .357 Magnums, tear gas, gas masks - with Marine Security Guards at the forefront - but soon Saddam's force would withdraw from the embassy wall.

Embassy foreign nationals reported that the Kuwaiti guards had literally stripped off their uniforms and fled when they spotted the Iraqis on the morning of 2 August. We were allowed to proceed, and the anxiety did not subside until we crossed over the steel barrier gate and into the enclosed parking lot. The embassy security officer informed us that we would be billeted in the Marine House. This was the small complex used as the living compound for the five Marine guards stationed at the embassy. They were now living and working full-time in the Chancellery building, since some twenty families now occupied their quarters.

On the evening of 8 August we got an ugly reminder that the Iraqis had us surrounded. At 2045 hrs automatic weapons fire began arching over the embassy compound. The Marine guards immediately alerted everyone to head for the Chancellery vault. Apparently, the embassy was caught in a crossfire between Iraqis and Kuwaiti resistance fighters.

At one point the Ambassador ordered the Marine Guards out of uniform and instructed the embassy security officer Chip Bender to destroy all weapons and the USLOK secure communications.

On September 14 Iraqi soldiers stormed the French, Belgian and Canadian diplomatic buildings in Kuwait and briefly detained five diplomats, including a U.S. consul. France responded by announcing that it would send four thousand more soldiers to the Persian Gulf and by expelling Iraqi military attaches in Paris. Throughout it all, the American Embassy remained secure.

KNOW THY ENEMY

MSG Det Lima, Peru

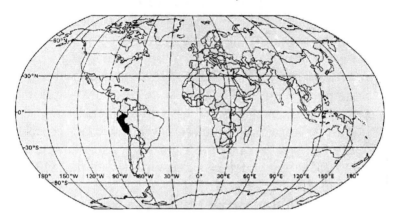

It is important to study the threats in the country to which you are assigned. You never know when that knowledge may come in handy – or save your life.

Lima Peru is one of the most beautiful posts in South America, with much to offer MSGs posted there – but it is also the capital of one of the most violent countries in the Western Hemisphere. Leftist rebels have attacked the United States Embassy there with dynamite, RPGs and submachine guns on numerous occasions over the past twenty years, and that does not appear likely to stop in the near future.

On one occasion dynamite was thrown from a passing car and exploded at the front gate. On another an RPG penetrated the compound and killed several FSN guards. Yet another attack occurred when the vehicle carrying MSGs was unsuccessfully targeted in a rocket attack. All occurred without warning – and every one of them should have been anticipated.

The rebel groups 'Tupac Amaru' and 'Sendero Luminoso' (Shining Path) are responsible for these attacks, and are among the most ruthless terrorist organizations in the world. On December 17, 1996 fourteen members of the Túpac Amaru Revolutionary Movement (MRTA) took hostage hundreds of high-level diplomats, government and military officials and business executives who were attending a party at the official residence of Japan's Ambassador to Peru in celebration of Emperor Akihito's birthday. Most of the hostages were soon released, and after being held hostage for 126 days the remaining dignitaries were freed on 22 April 1997 in a raid by Peruvian Armed Forces commandos during which one hostage, two commandos and all the MRTA militants died.

In recent years we have seen many more attacks, most significantly in 2002 when Shining Path terrorists detonated a powerful car bomb which killed nine people and injured thirty near the U.S. Embassy in Lima in advance of a visit by President George W. Bush.

All of the traditional anti-terrorist strategies naturally apply in such a place, but it would be wise for diplomats and MSGs alike to familiarize themselves with those organizations, and with the underlying causes of these incidents, in order to be better prepared. They key to surviving is not so much how you react to an attack, as it is in knowing how to avoid one.

As the Watch Commander on *Hills Street Blues* used to say to his cops as they prepared to hit the streets, "Be careful out there!"

STILL FLYING
High and Bright

MSG Det Manama, Bahrain

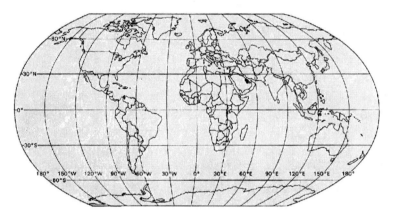

This first person account shows what can happen at a quiet, "friendly" post on a moment's notice, and gives some insight into the "spin control" exercised by some governments in the wake of such incidents.

The Embassy in Bahrain was attacked at 4:00 PM, and the five Marine Security Guards and two Regional Security Officers were the only ones protecting it. The Embassy was closed, so luckily we only had to protect ourselves and the compound from destruction. The protest was a pro-Palestinian rally that had started a few miles away and became violent as radical Muslim groups took charge and led the march on the embassy at 3:30 PM. There were between 2600 and 3000 protestors, but we do not have an exact count because it just happened several hours ago. The protestors were throwing rocks, sticks and fire bombs at the embassy, and between twenty and thirty protestors climbed the wall and started setting cars, trucks and the embassy itself on fire. When the protestors climbed the walls, the five Marines were authorized to use necessary force. We shot tear gas into the crowd to try and turn them away from the compound, but they continued the destruction of our POVs and satellite communication equipment - which was outside the embassy, but inside the compound wall. At 4:10 PM no firefighters had arrived, so three Marines had to leave the embassy under covering fire of the Marines on the third floor to go to the back of the compound to put out the fires themselves. We sustained minor injuries from the protestors attacking us with rocks and sticks as we chased them out of the compound and back over the walls, while the remaining Marine put out the fires as best he could. After the fires were extinguished the Marines pulled back into the embassy and continued to force the protestors back onto the highway behind the embassy. The protests ended by 7:00 PM, and the investigation was in full swing by this time as well. The Marines stayed at the embassy for security throughout the night and will continue to do so today. The host government was embarrassed that they could not control their own people, and that their riot control police were no match for the number of protestors.

The King of Bahrain instructed the Ministry of Interior to fix the problem before dawn, so that minimal damage would be seen this morning. When the Marines were told at 6:00 AM this morning we could leave to go get some sleep, we drove past the outside of the embassy and were amazed. The graffiti had been painted over, the guard shacks which had been burned were replaced, the fire bomb burns on the wall had been painted over, the thousands of rocks that had been thrown were swept away, and the barbed wire which had surrounded the embassy the previous day was gone. Other than the total destruction of all government property inside the compound, it looked to anyone passing by like nothing had really happened.

The host government representatives called the Ambassador last night and told him they would replace everything that was destroyed, and that cost was not an issue. The Kingdom of Bahrain is so scared of pissing off the U.S. that it will do anything in its power to preserve its relationship with us - even in the shadow of the ignorance of their own people.

The entire fight was caught on video, because we saw the Al Jazerra News choppers over our heads. The video may or may not be released, because it will no doubt be handled the same way the Daniel Pearl murder was handled - the government will try and influence what is released. As a matter of fact, the news agency that filmed the riot yesterday is the same one that had the Daniel Pearl murder on tape.

This is the story of what took place on April 5th 2002 in Manama, Bahrain - regardless of what you see or hear. Before I go treat my scrapes and cuts and then go to bed for about twelve hours, I would like to say that the two RSOs and five Marines kept the Embassy from burning into an embarrassing pile of the United States' image, and successfully held off (and sometimes injured) anyone who intended to come take our pride from us.

Well, I am off to bed because I am dead tired, but it was worth it because I had the time of my life yesterday - by far the most fun I have ever had. And one more thing - the American flag is still flying as high and bright as it was before the riot, while the protestors' flags are lying on the ground, covered in mud!

THE DEPARTMENT OF STATE

VISITOR ON DECK

MSG Detachment Bangkok, Thailand

Chris Tuvell

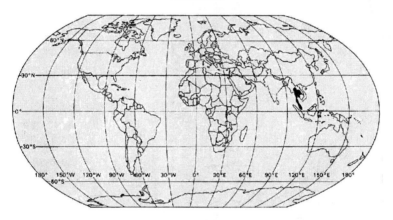

MSGs are expected to be aware of everything going on within the embassy compound for obvious reasons, but the State Department doesn't always remember to notify them. The moral of this story is twofold - strange things can and do happen on the midwatch, and you never know <u>who</u> you might meet on the MSG program!

My second posting was at the American Embassy in Bangkok, Thailand, where I was stationed from July 4th 1984 until November 25, 1985. The Marines there had three posts to man during the week. Post One was at the front of the Embassy, Post Two was a roving patrol around the embassy compound (roving to eleven outlaying agencies on mids) and Post Three was in the Consular Section of the Embassy (during the day only). I can remember clear as day one night while on Post Two during mids, coming back from checking ten of the eleven places I had to check. The last place on the list was the Ambassador's residence. It consisted of the main residence, and a two-story visitor's cottage located near the front gate.

The visitor's cottage was usually dark during my routine patrols on Post Two, but on this night the house was very bright. It was 3:30 in the morning, and all of the lights were on in the cottage. Thinking this highly unusual, I stopped to check and ensure everything was okay. I went throughout the entire first floor, which consisted of four or five rooms, and found all of the room unlocked and empty. I proceeded upstairs and checked two of the three rooms which were located on that floor, and both were unlocked and empty. I proceeded to third and final room, but the door was locked. I knocked on the door, but the room was silent. I waited a few seconds and knocked on the door again. In a groggy voice a man said, "Who is it?" I answered back, "Marine Guard, sir. May I please see some identification?" and he answered back, "Just a second." When he answered the door, I again said, "I'm sorry to bother you at this hour sir, but the house is usually dark - but tonight the lights were all on and I want to ensure everything is alright." He handed me his State Department ID and said,

"It's alright, I know you're just doing your job." I handed him back his ID and said goodnight, and thank you. I finished my patrol and went back to the embassy. At 6:15 AM I went out to raise the colors, and upon my return the daily English paper had arrived and the cover pager had a picture of the gentleman I had woken up earlier at the Ambassador's visitor cottage. The headline read: "AMBASSADOR TO THE UNITED NATIONS, VERNON WALTERS, IN TOWN." I freaked out for a moment, and explained to my fellow watchstander on Post One the person on the cover was the same person I had woken up in the middle of the night. We had a good laugh about the whole incident - and both of us were glad I had handled the situation the way I did!

A CHANGE OF HEART

MSG Detachment Brazzaville, Congo

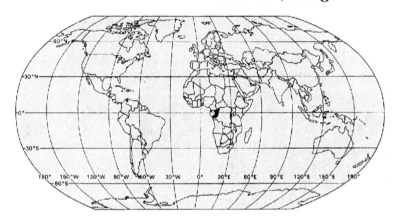

It's funny how a person's views can be altered by the reality of a situation, isn't it? I equate "dovish" State Department employees who consider Marines to be violent brutes with an antigun activist whose family is being threatened by an intruder – and suddenly loses their aversion to firearms.

Politics are part and parcel of serving in the State Department. There's no getting away from it. We in the military, on the other hand, are trained to avoid becoming embroiled in politics - but sometimes you just have to speak up for what you believe.

As civil war heated up in the Congo there were no safe places in the city of Brazzaville, but some were less safe than others. Stray bullets would come through walls without warning, and there were roadblocks manned by armed soldiers everywhere. Even the Ambassador was in the line of fire, and one night he had a really close call. A firefight broke out near the Ambassador's residence, and Mr. Ramsey and his wife had to take cover in an interior hallway. They waited there for several hours with nothing to drink but a bottle of champagne, which he had grabbed in a daring dash to the kitchen. When things calmed down some of us went down there to move Mr. and Mrs. Ramsey to an apartment near the Embassy, and upon entering the compound found that the brand new Jeep Cherokee he had shipped into the country was shot full of holes. I was glad it was just the car, and not the Ambassador!

This compound in Brazzaville near the author's residence suffered several direct RPG hits during the fighting.

The Administrative Officer of our Embassy at that time was a woman named "Sandy." She was third in the chain of command, but nothing at all like the Ambassador. Sandy despised the military in general, and Marines in particular.

She had attended Berkley back in the 1960's, and as you might suspect her politics were just to the left of Fidel Castro. I did my best to avoid confrontations with her, but she had a habit of taunting me about my politics every opportunity she got. As you may have guessed, we were *not* friends.

Sandy lived on the edge of the Bacongo district of Brazzaville, which is precisely where government control of the city ended and rebel control began. One fine day it was her turn to get caught in the crossfire, and things quickly went from bad to worse. Rounds from Soviet-made 12.7 heavy machineguns were tearing through the walls of her house like they were made of paper mâché, so Sandy lay down on the floor and radioed the Embassy for help.

When the call came in the Regional Security Officer asked me if I wanted to "go for a ride." He didn't tell me where we were going at first, probably because he thought I might join the rebels who were shooting at Sandy. We armed ourselves and headed for Bacongo in a hardened Suburban that belonged to the Defense Attaché - it wasn't an armored vehicle, but did have ballistic glass, a reinforced body, and self-sealing tires.

When we arrived at the gate to Sandy's residence we could see some rebel positions on the far side of the ravine, and there was an armored vehicle of the Congolese Army firing at them from a spot just up the road. We went in and grabbed Sandy as quickly as we could, threw her in the Suburban, and made tracks. As we headed toward home Sandy suddenly blurted out, "I've never been so happy to see Marines in my life!"

If there is one thing I hate it's a hypocrite, and if looks could kill Sandy would have died then and there.

DICK DINGLEHOFFER

MSG Detachment Kinshasa, Zaire

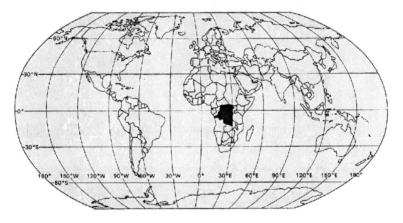

Some people in important positions are oblivious to obvious dangers, while others make like Chicken Little and scream that the sky is falling at the smallest provocation. The jury is still out on which is worse.

During my time as a Detachment Commander on the Marine Security Guard Program I was required to work closely with members of the State Department, and it was an eye opening experience to be sure. I had always expected the people entrusted with formulating and implementing the foreign policy of the United States to be something special, and while there were a *few* who lived up to my expectations most of the members of State I knew were hypocritical, selfish prima donnas whose only interest was the advancement of their own careers. Whenever a decision had to be made their first thought would invariably be "how will

I look" rather than "what is the right thing to do."

One of the Foreign Service's all stars of incompetence was the Regional Security Officer in Kinshasa, Zaire, who I will call Dick Dinglehoffer. He was such a buffoon that virtually everyone in the Embassy, including his very own assistant RSO, invariably referred to him by using a number of decidedly unflattering nicknames when not in his presence. Dick had spent many years developing his reputation, with his crowning achievement being his inept handling of security at the Embassy in Moscow where he was largely responsible for the now infamous Lonetree incident. Why he wasn't sacked remains a mystery to this day, as anyone who is familiar with Diplomatic Security and has read the book *Moscow Station* will agree.

The Stanley Pool portion of the Congo River, which separates Brazzaville from Kinshasa, Zaire (now the Democratic Republic of the Congo).

I am not inclined to believe a lot of what I hear because facts tend to get distorted in telling and retelling a story, so I gave Mr. Dinglehoffer the benefit of the doubt until I had the opportunity to judge for myself. Dick, for his part, wasted little time proving his reputation was not only well deserved, but had in fact been grossly understated.

Zaire, while not the most stable country in the world, was no worse than most of the other so called African republics. Rife with disease, inflation (One million Zaires equaled about twelve U.S. *cents* in 1993) and corruption, the country had been under the rule of the dictator Mobutu since gaining independence from Belgium some thirty years earlier - so it was not the least bit surprising when civil war broke out.

The main reason we visited our Embassy across the river in Zaire was they had a small commissary, and we didn't. The ferry between the two countries had stopped running due to the ongoing conflict, so it was decided I would cross the river with one of my MSGs in the Embassy boat to pick up supplies for our families. Easier said than done. What started out as a simple shopping trip turned into an odyssey filled with international intrigue.

Martial law was proclaimed in Zaire just about the time I shoved off from Brazzaville - and if I had known I wouldn't have gone. When I arrived the Zarois customs people kept my diplomatic passport, no doubt in an attempt to elicit a bribe in exchange for its return - and that just wasn't going to happen. I was told by the Kinshasa RSO to stay in Zaire until it was returned, but as time dragged on I grew more and more impatient. My detachment in the Congo was "down two," and the guy telling me it was "too dangerous to cross the river" had recently declared a state of emergency because someone had heard a tire blowout. Dinglehoffer apparently thought it was a gunshot.

Since the RSO had deemed a river crossing to be too hazardous, a plan was hatched to smuggle us out on a small commercial airplane. We were each issued new civilian passports, but were prevented from boarding because they did not have an entry stamp. The Zarois Army arrested us as spies, and took us to a holding area. I quickly came to the conclusion that NOW was the time for bribery, and slipped some money to a guard in exchange for our freedom.

A couple of days later I decided enough was enough. I had the detachment chauffeur drive my corporal and I down to the Embassy dock, and we cast off before anyone could stop us. We crossed the Congo River without incident, and the dangers the RSO had warned us about never materialized. Naturally Dinglehoffer raised hell when he found out we had left, and of course that led to an investigation.

My CO sent an investigating officer out from Nairobi to check into the RSO's complaint, but things worked out for the best in the end. The Captain who came thought Dinglehoffer was a moron, and understood why I had done what I had done. What's more, he had served in the Gulf as a platoon commander with 1st Force Recon, and we had a lot of mutual acquaintances. Of course that had no bearing whatsoever on his findings…

That same officer eventually part of the Marine Corps Congressional liaison office after his tour on the MSG program was over, and several years later accompanied a delegation of politicians to Camp Pendleton for a "dog and pony show." I was manning an exhibit of special operations radio gear, and had no idea he was in the AO. He was now a Major, and when he reached my display he took one look at me and exclaimed, "Holy shit. I heard you were dead!"

Fortunately, that report of my demise was somewhat premature.

BART BONEHEAD

MSG Detachment Brazzaville, Congo

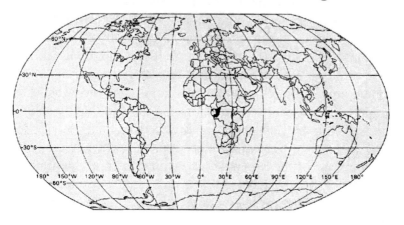

One of the most important things a Detachment Commander must do is forge a good working relationship with the RSO at his embassy – but sometimes that is just impossible!

One of the biggest problems with the Marine Security Guard Program is the requirement for experienced, professional Staff Non-Commissioned Officers with years of formal leadership training to answer to a series of State Department "Officers" who have none. They tend to think their lofty positions are all that is required to command respect, failing to realize it is a precious commodity which must be earned. That, along with the requirement to satisfy the needs of two parallel chains of command simultaneously, makes the job of a Detachment Commander far more complicated than it need be.

An example of such an incompetent, self-serving State

Department employee was an individual who I will call Bart Bonehead. Mr. Bonehead would be a perfect subject for a Psych 101 class because his problems are so transparent that they would be easily identifiable to every student. A short, fat, one-eyed little troll, Bart no doubt became a Special Agent in the Diplomatic Security Service in an attempt to attain some degree of the manhood he had been denied by genetics – and in the process opened to question the entrance standards of DSS. He was such an insecure individual, and possessed so little personality, that he felt it necessary to create an artificial persona for himself in an attempt to impress others with his manliness. He thought that by calling himself Cowboy and wearing a collection of pointy boots and ratty cowboy hats he would be macho, when in reality he became one of those people we've all known at one time or another who give themselves a nickname because no one else will. As a result, he became a bit of a laughingstock in the process.

This building in Brazzaville served as the American Embassy during the author's tour of duty there. It had previously been a bank, and was filled with many odd nooks and crannies.

When Mr. Bonehead was first assigned to the American Embassy in the Congo he professed to be a big supporter of Marine Detachments and was initially welcomed with open arms. It wasn't until later that we figured out he liked to hang with the Marines because he figured by associating with and dictating policy to the tough guys at post he would be a tough guy too. He was actually assigned to Brazzaville as the Administrative Officer on an "excursion tour" outside of the security side of the house because he ostensibly had hopes of one day becoming an Ambassador, an aspiration which even his friends considered to be laughable given his low self esteem, mediocre skills and lack of style.

During my tenure as Detachment Commander I had numerous occasions to butt heads with this walking advertisement for legalized abortion, and it is a testament to my self-control that he remained ambulatory throughout our six months together. It is because of people like him the Corps needs to review the agreement which assigns Marines to Diplomatic posts in support of the State Department. Too many decisions concerning the Marines are left to the discretion of the civilians working for State, and too often the Detachment is treated like a group of second class citizens. It is impractical to expect Marines to operate efficiently when they are subject to the whims of people who have no concept of professionalism, loyalty and integrity. What invariably happens is a confrontation the MSGs cannot possibly win occurs whenever the Marines do not automatically bow to the wishes of the State "officer" who happens to be in control at the moment.

The bottom line is I'd love to have the used car salesman who sold this concept to the Marine Corps working for me, since he could probably sell a glass of water to a drowning man!

THE PANIC BUTTON

MSG Detachment Brazzaville, Congo

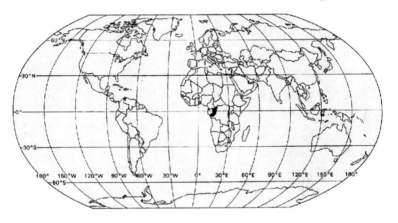

A good RSO is worth his weight in gold, and will make a Marine's time at post go by both quickly and productively. Despite my problems with "Mr. Bonehead" I actually did pretty well at both of my posts in that regard. This story is dedicated to Jim Schnaible, Jim Ennis and John Eustace, three of the best in the business. It's too bad they can't all be like those guys!

When I first assumed command of the Marine Detachment in Brazzaville, the State Department's Regional Security Officer there was a fellow by the name of Jim Schnaible. It was my first post and I wanted to hit the ground running, so once I had gotten settled in one of my first official duties was to go around with Jim to the homes and compounds belonging to the Embassy and assist him in conducting a security survey. That consisted of evaluating such things as compound walls, lighting, guard posts and alarm systems - and we had had an extensive checklist to make sure we didn't miss anything.

The last residence to be checked was the Marine House where my detachment resided, and the last item on the checklist was the alarm system "panic button." I had seen these in the other residences, but wasn't yet familiar enough with the Marines' quarters to know where it was located. While Jim waited, I called the Marine on duty at Post One to find out where it was, and the moment I hung up the phone the RSO appeared in front of me with pen and clipboard in hand and an expectant look on his face.

"Well?" he asked.

"There isn't one," I replied.

"What do you mean?"

"There is no panic button in the Marine House."

After letting that sink in for a few moments Jim just shrugged his shoulders and wrote in the space allotted, "N/A – Marines don't panic!"

Ambassadors In Blue

INTERNATIONAL RELATIONS

TO HONOR MY COUNTRY

MSG Detachment Unknown

Anonymous

A foreign diplomat who often criticized American policy once observed a United States Marine performing the evening colors ceremony at an American Embassy overseas. The diplomat wrote about that simple but solemn ceremony in a letter to his country, and it is because of such things that MSGs must always assume they are being watched by someone, and perform their duties with great skill – and pride.

229

"During one of the past few days I had occasion to visit the U.S. Embassy in our capital after official working hours. I arrived at a quarter to six, and was met by the Marine on guard at the entrance to the Chancery. He asked if I would mind waiting while he lowered the two American flags at the Embassy. What I witnessed over the next ten minutes so impressed me that I am now led to make this occurrence a part of my ongoing record of this distressing era.

The Marine was dressed in a uniform which was spotless and neat. He walked with a measured tread from the entrance of the Chancery to the stainless steel flagpole before the Embassy, and almost reverently lowered the flag to the level of his reach where he began to fold it in military fashion. He then released the flag from the clasps attaching it to the rope, stepped back from the pole, made an about face, and carried the flag between his hands - one above, one below - and placed it securely on a stand before the Chancery. He then marched over to the second flagpole and repeated the same lonesome ceremony.

On the way between poles, he mentioned to me very briefly that he would soon be finished. After completing his task he apologized for the delay - out of pure courtesy - as nothing less than incapacity would have prevented him from fulfilling his goal. He said to me, 'Thank you for waiting, Sir. I had to pay honor to my country.'

I have to tell this story because there was something impressive about a lone Marine carrying out a ceremonial task which obviously meant very much to him and which, in its simplicity, made the might, the power and the glory of the United States of America stand forth in a way that a mighty wave of military aircraft, the passage of a super-carrier, or a parade of ten thousand men could never have made manifest.

In spite of all the many things that I can say negatively

about the United States I do not think there is a soldier, yea, even a private citizen, who could feel as proud about our country today as that Marine does for his. One day it is my hope to visit one of our Embassies in a far-away place and see a soldier fold our flag and turn to a stranger and say, 'I am sorry for the delay, Sir. I had to honor my country'."

VIVE LA FRANCE!

MSG Detachment Algiers, Algeria

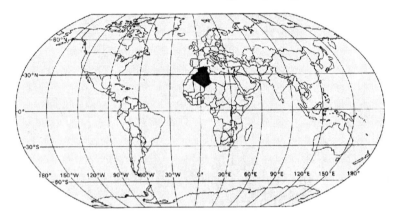

I concluded long ago that the art of diplomacy is something best left to diplomats, because Marines tend to tell the unvarnished truth to any and all who care to listen.

The French have always been something of an enigma to me. France is the country that gave us Napoleon, a man whose military exploits and thirst for conquest were legendary - yet during the past century that same nation has fielded some of the most inept armies in history, and has depended on other nations to save their bacon on more than one occasion. So it boggled my mind when the French opposed a proposed United Nations resolution to oust Saddam Hussein for not complying with a directive to disarm, going so far as to threaten a veto in the Security Council. After all, they certainly had no objection when we sent troops to liberate *their* country during both World Wars, did they?

World War I was initially fought by the Army, but once Marines arrived in Europe they quickly distinguished themselves. The battle for Belleau Wood became one of the most famous in Marine Corps history. The 5th and 6th Marine Regiments distinguished themselves with such valor the French renamed the area "Forest of the Marines," and it was there we earned the nickname Devil Dogs.

The Marine Corps didn't fight in Europe at all during WWII, but I had a personal stake in the liberation of France nonetheless. My own father landed on the beaches of Normandy on D-Day in June of 1944. Tens of thousands of Americans died driving the Germans from French soil, but fortunately my Dad was not one of them. If he had been, I wouldn't be here to write these words!

My feelings for the French can be best illustrated by something that occurred a few years back. A Marine colonel who happened to command MSG Battalion was attending a state dinner in a former Francophone colony in North Africa on the occasion of Bastille Day, and naturally his Sergeant Major accompanied him. As per protocol the Sergeant Major was seated next to the French military attaché, who was also a colonel. As the evening wore on and the libations flowed (the Sergeant Major was not a temperate man) the Marine began eyeing the Frenchman with a growing amount of disdain. Finally, when he could stand it no longer, he turned and asked a question many of us have probably wondered about.

"So tell me Colonel, how many French troops does it take to defend Paris?" he asked.

After thinking about it for a long moment the Frenchman replied "I do not know, monsieur."

The Sergeant Major smiled with thinly veiled glee. His

prey had taken the bait.

"Of course you don't. It's never been done!" He was of course referring to the French habit of declaring Paris an open city in the face of advancing German forces. The French Colonel was dumbstruck by the comment, and the Marine couldn't resist twisting the knife a bit more. "But don't you worry, we'll always be happy to come on over and take it back for you *anytime*."

WHAT A MESS

MSG Detachment Canberra, Australia

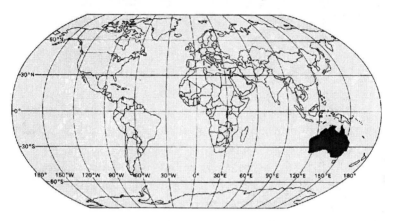

Building a good rapport with the other services who may be represented at a diplomatic post, as well as with the host nation's military, can go a long way toward the success of an MSG Detachment – and the employment of alcohol can be most helpful in making that happen!

If you have never attended a Mess Night, or worse yet have never heard of such an occasion, you are missing out on one of the finest traditions the Marine Corps has to offer. It is an evening of camaraderie that, when properly run, can evoke many emotions in even the hardest of men. One of the best accounts of a Mess Night I have ever read is contained in the book *The Great Santini* by Pat Conroy. I suggest you read it.

The Marine Corps Mess night is an occasion which has its roots in the British practice of a formal gathering of men in an atmosphere which contributes to the unity and esprit of an organization. The uniform for such an event is Mess Dress, blues, or black tie for civilian guests. The affair is always stag, and is presided over by the President of the Mess. His word is final on all things, and there is no appeal.

The evening begins with a cocktail hour, followed by a formal dinner. When it is time to serve the main course the chief steward "parades the beef" to the head table, accompanied by a drummer and fifer playing "The Roast Beef of Olde England." Once the President pronounces it "tasty and fit for human consumption," dinner is served.

Once the dinner has concluded port wine is passed to each member of the mess, and formal toasts are proposed. That is followed by a final "bottoms up" toast with traditional 1775 rum punch to the words "Long live the United States, and success to the Marines." At that point the mess is adjourned to the bar, and the evening progresses in accordance with the impulse and ingenuity of the members.

There are many rules which must be observed within the mess, not the least of which is a prohibition on the topics of politics, religion and women. "Charges" of improper behavior may be brought against any member of the mess, save the President and Guest of honor, by any other member. Proper etiquette is enforced through the practice of fining, and the imposition of such fines can be quite enjoyable to watch when done with a bit of humor. All fines collected go on the bar at the conclusion of the evening to defray the cost of after dinner drinks.

I have been fortunate in that I have attended numerous such occasions, and even had the privilege of hosting a few. One particularly enjoyable Mess Night took place in Canberra and was attended by representatives of all services and several nations. As President it was my duty to make some opening remarks, and I didn't want to spout off with the same old politically correct drivel that tends to be the norm at international functions. In preparing my remarks I realized it is a well known fact Australia was originally a penal colony for the British, and thought that would be an appropriate subject to begin the evening with.

"When I applied for my visa to come to Australia I was asked if I had a criminal record," I began. "Now, I'm aware of your history, but I didn't realize it was still a requirement for admittance!"

From that point on the evening just got better and better. But it is not *all* fun and games.

An important part of the Mess Night tradition is the Fallen Comrades table. No one sits there. It is symbolic of our comrades who have fallen in battle, and is covered in a black tablecloth and adorned with the cover, gloves and sword of a Marine - lest we forget.

TOYS FOR TOTS

MSG Detachments Brazzaville, Congo and Canberra, Australia

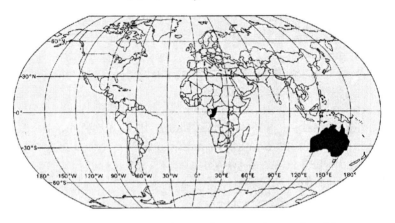

The title of this book, 'Ambassadors in Blue,' should serve as a clue regarding the secondary "mission" of Marine Security Guards. Our main function is obviously to provide for the security of diplomatic posts and personnel, but it is also important to remember we are one of the most visible symbols of our country overseas – and the Toys For Tots program is one of the best ways to show them what we are all about!

Toys for Tots began in 1947 when Major Bill Hendricks, USMCR and a group of Marine Reservists in Los Angeles collected and distributed five thousand toys to needy children. The idea came from Bill's wife, Diane. In the fall of 1947 Diane handcrafted a Raggedy Ann doll and asked Bill to deliver it to an organization which gave toys to needy children at Christmas. When he determined that no such agency existed, Diane told Bill that he should start one. He did. The 1947 campaign was so successful that the Marine Corps adopted Toys for Tots in 1948 and expanded it into a nationwide campaign. That year, Marine Corps Reserve units across the nation conducted Toys for Tots campaigns in each community in which a Reserve Center was located. Marines have conducted successful nationwide campaigns at Christmas each year ever since.

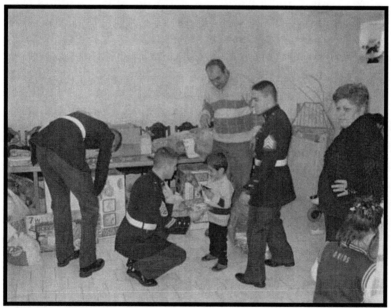

Toys For Toys is a great way to help kids and show the State Department that Marines are more than just steely-eyed killers.

Bill Hendricks, a Marine Reservists on weekends, was in civilian life the Director of Public Relations for Warner Brothers Studio. This enabled him to convince a vast array of celebrities to provide their support. In 1948 Walt Disney designed the Toys for Tots logo, which we still use today.

Disney also designed the first Toys for Tots poster used to promote the nationwide program. Nat "King" Cole, Peggy Lee, and Vic Damone recorded the Toys for Tots theme in 1956, and Bob Hope, John Wayne, Doris Day, Tim Allen and Kenny Rogers are but a few of the long list of celebrities who have given their time and talent to promote this worthy charity. First Lady Barbara Bush served as the national spokesperson in 1992, and in her autobiography named Toys for Tots as one of her favorite charities.

My own affiliation with the Toys for Tots program began during my days with 2nd Battalion, 25th Marines in Garden City, New York. Collection points were set up at Shea Stadium prior to a Jets football game one Christmas season, and volunteers were needed to man them. I admit my motive wasn't completely noble that day - participants were allowed into the stadium to watch the second half for free - but the experience did teach me a lot about the spirit of giving.

Once I returned to active duty deployments and operational tempo often limited my participation in the program - but I still donated toys whenever possible. When I reported for duty in the Congo during the later part of 1992 one of the first things that struck me was the overwhelming poverty of the Congolese people. I wanted to help in some small way, so my Marines placed a large box under the Embassy Christmas Tree and sent a memo to every office soliciting the donation of toys. Everyone was very generous, and what we collected was distributed to local children by missionaries working in the area.

My next assignment was in Australia. While there was nothing like the poverty I saw in the Congo, there were still a lot of needy kids all the same. Once again a collection point was set up under the Embassy Christmas Tree, and once again we were overwhelmed by the generosity of everyone who worked there. Since there was no official Toys for Tots distribution system there we turned the toys over to the Australian Salvation Army. They in turn brought them to children's hospitals where the neediest children received an unexpected visit from Santa Claus.

No story about Toys for Tots would be complete without mentioning my friend Sam Dipoto, who ran the program for the Marine Corps League in the Tampa/St. Petersburg area of Florida for nearly twenty years. Sam loved to tell the story of a woman who donated a dozen brand new bicycles one Christmas. When he tried to thank her she said it was just her way of thanking Toys for Tots. It turned out a couple of years earlier the woman had been out of work and destitute, but the Marines had come through with a bicycle for her young son.

I have no doubt that there are many like Sam all across the country (and around the world), and hopefully there will be many more to follow in their footsteps. The kids are counting on it.

If you would like to donate a toy, make a cash contribution or donate your time contact the Marine Toys For Tots Foundation, PO Box 1947 Quantico, VA 22134. They can also be reached at (703) 640-9433 or via the worldwide web by visiting www.toysfortots.org

THE KING OF TONGA

MSG Detachment Canberra, Australia

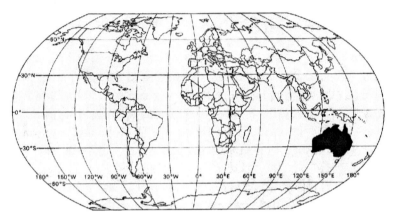

Most of the stories in the book are included in order to show MSGs in action around the world, or to make a point about operational matters – but this is not one of them. The point of this parable is, "Know when to keep your mouth shut!"

Marine Mess Nights are in many ways unique, but we do not have a monopoly on such occasions. Australians, as well as other Commonwealth nations, have a similar occasion known as the Dining In. While it is different from our Mess Night in a number or ways, the protocol is very similar.

I was attending such a dinner at the Royal Military College in Canberra Australia when I was introduced to the new Regimental Sergeant Major, a fellow named Jack Selmes. Jack was quite the professional soldier. In fact he looked so much like a crusty Sergeant Major he was practically a caricature.

During the evening a few of the Aussies tried to convince me to "ask the RSM who the King of Tonga is." I thought better of it and asked around instead, and was glad I did. I had learned never to ask questions at the behest of others back when I was a slick-sleeve Private. One of my new "friends" had encouraged me to ask another Marine why he didn't buy his sister a bicycle for Christmas, and when he answered that she didn't have any legs (it wasn't true) I felt about six inches tall. It was a cruel joke, but it's one I have never forgotten. Fool me once shame on you, fool me twice shame on me.

I soon discovered the RSM had once been banished to some remote posting due to an incident which had occurred during a Dining In some years before - and that he was very touchy about it. The story goes that Jack had been imbibing rather heavily prior to the dinner, arrived late, and stumbled to his place at the table just as the first toast was made. Keep in mind it is customary to toast the head of state of every country in attendance, followed by the "loyalty toast" to the Queen. Just as Jack took his place the President of the Mess raised his glass and proclaimed "the King of Tonga!" Jack had no idea there were Officers from that island nation present, and he had quite possibly never heard of Tonga, period. He belched loudly and in his semi-inebriated state blurted out, "who in the *hell* is the King of Tonga?" during one of those embarrassing moments of total silence which sometimes occur in a room full of people.

Unfortunately for Jack, one of those silent people was a very unamused Australian General Officer.

LE GRANDE CHEF

MSG Detachment Brazzaville, Congo

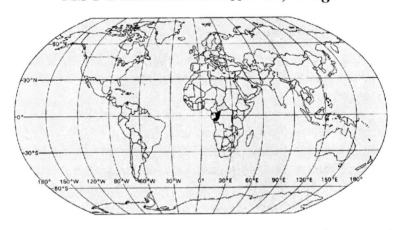

Sometimes you can get by on your reputation alone – and fortunately, <u>this</u> was one of those times. I want to extend a personal thank you to Chesty Puller, Dan Day, John Basilone and all of the other Marines who built the reputation of our Corps!

I have never had an aptitude for languages. Just ask my high school German teacher. I did fairly well with the vocabulary, but when it came to conjugating verbs and all that stuff I was lost - and I figured "who needed it" anyway…

When I first arrived in Brazzaville I encountered a number of roadblocks manned by the Congolese Army, and on more than one occasion thought some drunken and undisciplined soldier was about to end my life with a rusty AK-47. It was at that point I decided it would be prudent to learn the local lingo, in this case French. While most language students start out with phrases like "my name is John," I memorized things more along the lines of "please don't shoot!"

On the 4th of July in 1993 I celebrated Independence Day with some British friends, which I found to be quite ironic when you consider they were our opponent during the Revolutionary War. The Diamond Boys, as they were known by us, were a group of ex-pats living in the area, and as you may have already guessed they worked as quasi-legal diamond smugglers in West Africa.

Richard of the "Diamond Boys" on a sandbar in the Congo River. The occasion was the marriage of one of the author's MSGs to a local Belgian girl.

Our outing was to be a picnic on a sandbar in the middle of the Congo River, and our mode of transportation was a speedboat the Brits used to move their "inventory" across that same body of water. There were seven of us in all, including one of my Marines and his French girlfriend.

As we cruised down the river there were several bursts of automatic weapons fire in our general direction, but since gunfire was not unusual in Africa we ignored it and continued on to our destination. The picnic lasted three or four hours, and when we were finished playing "the Queen vs. the Colonies" we packed up and headed for home.

All was fine until boatload of Congolese soldiers armed with AK-47s and an RPK pulled alongside and ordered us to follow. We were taken by them to a remote island in the middle of the Congo River which served as some sort of military outpost. I decided to try out some of my rudimentary French in an effort to smooth things over, but from the reaction I got I must have insulted someone's mother. At that point I decided to let the French girl among us handle the negotiations.

After jabbering with them for a bit she confided in me that things were not going well. They knew who the boat belonged to, and wanted to know where the diamonds were hidden - but of course there *were* no diamonds. Just when I was beginning to go over my insurance coverage in my head the girl pointed to me and told them I was an American. They were unimpressed. She then added that I was a Marine, and their attitude changed visibly. I admit I was quite surprised these soldiers were familiar with U.S. Marines, because while the Corps certainly does have an unparalleled history of military accomplishments it wasn't the sort of thing you expect to be taught in Congolese history classes.

When she finally identified me as "Le Grande Chef," or big chief, of the Marine detachment those guys became downright friendly. They not only let us go, they even offered to let me try firing their weapons. Go figure.

G'DAY MATE

MSG Detachment Canberra, Australia

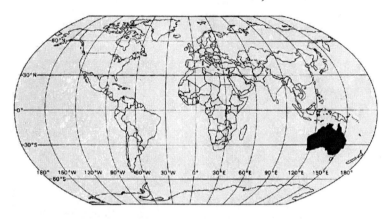

It is said that Australia and the United States are two countries separated by a common language – and that isn't far from the truth. I highly recommend that all MSGs learn the customs, and if possible the language, they will encounter at their next post – or suffer the consequences.

On my first day in Australia I received a rather expensive education in linguistics from the locals. I had just reported in as the new Detachment Commander at our Embassy in Canberra, and the Gunny I was replacing suggested the happy hour being held at the Royal Military College Sergeants Mess would be a perfect opportunity to meet my counterparts in the Australian Army. I still smelled like jet fumes from the long flight over, but deferred to his judgment and went along.

The Aussies turned out to be the salt of the earth. They

made me feel welcome from the very first moment, and bought me one Victoria Bitter after another. I didn't realize it at the time but a "VB", like most Aussie beers, has about twice the alcohol content of an American beer. I was having a blast until one of my new found friends said, "Hey Yank, how about shouting the bar?" It sounded like some sort of game to me, and in my semi-inebriated state I said, "Sure, why not?"

American Embassy Canberra was constructed to be representative of our Colonial heritage.

It turns out that "shout" means buy a round, and shouting the bar meant buying one drink for *each* of the hundred or so who were present. The only thing that saved me from abject poverty was the strong U.S. dollar, since the exchange rate was two-to-one in my favor. To their credit, I lost count of the number of times I heard one of those blokes say, "It's *my* shout, Gunny" over the next two years, and that more than balanced the books.

That wasn't my last lesson in Aussie lingo, not by a long shot. A couple of weeks later I was in a crowded downtown pub watching a rugby match on the "telly." I was beginning to like and understand the rough sport, but hadn't developed an allegiance to any particular team as of yet. In an effort to learn more about the game I struck up a conversation with the young lady standing next to me at the bar, and at one point casually asked who she was rooting for. She slapped me right across the face and stormed off without saying a word. Needless to say, I was a bit perplexed.

The next morning I related what had happened to a couple of the Marines who had been in country a bit longer than I, figuring they might be able to offer some sort of explanation for what had happened. They did - once they were able to stop laughing. "Root" is a versatile word which can be used as a noun, a verb, or an adjective, much like the "F-word" in the United States. I had unwittingly asked who she was sleeping with!

YOU JUST NEVER KNOW...

THE NEO

MSG Detachment Brazzaville, Congo

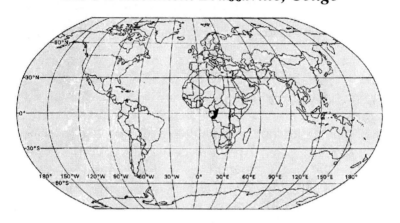

It's easy to find yourself between a rock and a hard place on the MSG program, especially when you are dealing with two different chains of command. I have it on good authority that the following incident was used by MSG School for years as an example of the sort of duties which do <u>not</u> fit the job description of an MSG. All I can say to that is, "Oh, <u>now</u> they tell me!" But I don't think it would have made a difference if they had, because in the end I was always guided by one simple question - "What would Chesty do?"

The Non-Combatant Evacuation Operation, or NEO for short, is one of the most important missions assigned to our military forces deployed around the world. Marines train exhaustively to perfect their ability to evacuate an entire expatriate American community from a foreign shore on a moment's notice, and we have seen the fruits of our labor time and again in places like Liberia, Somalia and Lebanon. One of the lesser known NEOs to be conducted by the Marine Corps occurred in Zaire during the early part of 1993 and regrettably never got a catchy code name like operations "Eagle Pull" and "Sharp Edge." That was probably because it was smaller in scope than most, two Marines to be exact, but it was a NEO nonetheless.

Brazzaville, Congo and Kinshasa, Zaire (now the Democratic Republic of the Congo) are not only the capitals of their respective countries, they are the closest neighboring capitals in the world, separated only by the Stanley Pool portion of the Congo River. Because of their proximity to each other and the volatile nature of the region it was inevitable for the two nations to constantly become entangled in each other's affairs, and sometimes we had no choice but to get involved also. Such was the case in the early part of 1993.

I had assumed command of the Marine Detachment in Brazzaville only a few months earlier, and had taken it upon myself to refit and repair the Embassy's twenty-five foot Boston Whaler. It was intended to be used to evacuate personnel in the event of an emergency, but the craft had fallen into disrepair due to a combination of abuse and neglect. Within a few weeks the fuel tanks had been purged, the outboard motors overhauled, and a two way radio was even installed. As it turned out the work was completed just in the nick of time.

Trouble had been brewing across the river in Zaire for weeks. President Mobutu, who had ruled the country since it gained independence from Belgium more than thirty years earlier, was a despot who also happened to be one of the richest men in the world. While the people of his country wallowed in poverty, Mobutu lived in opulence. One of the ways he maintained his lifestyle was by printing more money, which by the way bore his own likeness, whenever he was short on cash. While this certainly did fill his coffers, it also caused massive inflation and devalued Zarois currency to the point it was literally not worth the paper it was printed on.

The most common denomination was the one-*million* note, and a stack of them was required to buy even the most inexpensive item. Mobutu's solution was to issue the new five-million Z note, but when those new bills were used to pay the army all hell broke loose when shopkeepers refused to accept them. Many people were killed in the hours that followed, including a number of westerners.

Our Ambassador's wife, Luci Phillips, was unfortunate enough to be in Kinshasa when the fighting erupted. We all listened to her speaking with her husband on a hand-held radio while tracers and RPGs crisscrossed the night sky over the Zarois capital. It was clear by the tone of her voice that her situation was rather precarious. Then the following morning we learned the French Ambassador there had been shot and killed, and as the casualties across the river mounted it became clear she had to get out of there ASAP.

The Deputy Chief of Mission, Bill Gaines, approached me and asked if I would be willing to cross the river to pick up Mrs. Phillips. I agreed to do it, but stressed that I would have to get permission from my Company Commander first. In the meantime I had Corporal Grantham, who had volunteered to go along, prepare the boat while I attempted to place a phone call to Company headquarters in Nairobi. The Ambassador himself came down after about fifteen minutes and said that if we were going it was now or never. It was time to make a tough decision, and I made the one I believe any good Marine would make under similar circumstances.

Shortly after we launched the boat our Army Military Attaché, Lieutenant Colonel Jacobs, showed up at the dock with a truckload of French paratroopers. I hadn't expected that. They were armed to the teeth, and were being sent to reinforce their Embassy. What the hell, I figured, the more the merrier.

This is the Boston Whaler Corporal Grantham and I used in our infamous crossing of the Congo River into Zaire.

We weren't too sure what to expect when we reached the other side of the river, but within a few minutes the personal protection team from Kinshasa radioed instructions on where to land. The landing beach was located in a small cove, and I landed the French and withdrew as quickly as possible. There were snipers on the hill overlooking our position, and every few seconds I would hear the *crack* of a round passing overhead.

While the first boatload of paratroopers set up a defensive perimeter on the beachhead I returned across the river for some reinforcements. We landed a second time, and Mrs. Phillips was ushered aboard along with several other evacuees. For some reason the one that stands out in my mind is an Italian national who had been shot in the foot. I probably remember him because he was bleeding all over my nice, clean boat. I also noticed that the Ambassador's wife was in a bit of shock, but when I joking said, "Thank you for choosing Congo cruise lines" she laughed and seemed to snap out of it.

Detachment Commanders are required to send out what is known as an "Operational Incident Report" whenever something out of the ordinary occurs, and there was no doubt our little operation fit that profile. I sent the required message and as soon as the CO read it I received a scathing phone call. All I could think was, 'Oh, *now* the phones work!' It went something like this: "What the hell are you doing down there Gunny! Do you realize that you crossed an international border and for all intents and purposes *invaded* a sovereign nation?" I was in big trouble, at least until Mrs. Phillips wrote a letter praising all concerned for saving her life.

My CO, Colonel Duggan, was a big man in stature and even bigger in the ways that really matter. Once he had a chance to read Mrs. Phillips' letter and gave the matter a bit more thought he phoned to let me know he was calling off the dogs. Ever since that affair he liked to call me Gunny Highway, which I took as high praise – especially from him.

THE FINAL FRONTIER

MSG Detachment Canberra, Australia

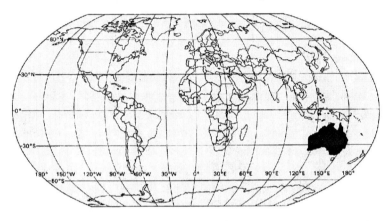

Sometimes strange and unexpected opportunities materialize out of thin air while you are on the program. My advice to MSGs is keep your eyes open, and your ear to the ground!

Anyone familiar with Marines knows that most of them are hams, and I guess I am no exception. When an opportunity to have our picture taken or get our name in the paper presents itself we are all "Hollywood" Marines at heart. I guess that's what prompted President Truman to once remark that a photographer was part of the T/O for every Marine rifle squad.

So I was quite pleased when the Air Attaché at our Embassy in Canberra, a Marine Major, asked me to field a phone call for him. The caller was a producer from the Fox television network, and he asked if I would be willing to work as a technical advisor on a movie about futuristic

Marines. Naturally I said yes, with the caveat I would have to get permission from my Company Commander.

The CO was a bit cool to the idea at first, since two of the Marines from the embassy in Bangkok had recently appeared in the Disney film *Dumbo Drop*. What had started out as a simple cameo turned into speaking roles, and each of the Marines had to fly back to LA to record voice-overs. I could understand the Colonel's reluctance to give the green light, and promised to keep things under control.

A couple of weeks later I was flown up to the Warner Studios on the Gold Coast of Australia, where my first task was to train a bunch of young actors to act like Marines. The first thing I noticed was every last one of them was in dire need of a haircut. I didn't expect high and tights, just something approaching regulation. I was quickly informed that the each actor's contract prohibited things which were considered by them to be "extreme." I suddenly understood why so many military movies were full of flaws you could drive a six-by through.

The star of the movie turned out to be R. Lee Ermey. He was well known to most Marines, having appeared in such films as *Full Metal Jacket*, *The Siege of Firebase Gloria*, and *Purple Hearts*. Ermy had in fact been a real Marine prior to his acting career, and was medically retired after being wounded in Vietnam. Like most Marines, I am a big fan.

When Ermey arrived at the hotel it was around dinnertime, and most of the cast and crew was gathered in the bar adjacent to the hotel's dining room. He appeared in the doorway, scanned the room for a few moments, picked me out from amongst the movie crowd, and then walked straight over to where I was standing. "You must be the Gunny!" he thundered. I guess I stuck out amongst all of those slimy civilians.

The author and R. Lee Ermey on the set of *Space* at the Australian RAAF base in Williamtown.

Ermy and I had dinner together in the hotel, during which I discovered that his on-screen persona is no act. When the waitress was slow to take our order he tactfully gained her attention by calling to her in the sweetest voice he could muster. "*EXCUSE* me, can we have some *SERVICE* over here!" I swear the poor girl jumped ten feet in the air, and from that moment on the service was great.

One morning after shooting had begun the producer pulled me aside and asked if I would be willing to play a small part in the movie. The actor they had contracted to play one of the assistant drill instructors was a bit overweight, and as a result didn't look very squared away in his uniform. I took one look at the individual in question and agreed he looked like ten pounds of manure stuffed into a five pound bag – there was no way I could allow him to portray a member of my beloved Corps. If "Gunny Hartman" had been there he would have no doubt told him "Son, you look like about 150 pounds of chewed bubble gum!"

The part was something I had been rehearsing for my entire life. I just had to run around and scream at people, and even though I had never been a Drill Instructor it was a piece of cake. Any questions I may have had about my believability were answered during a scene where our new recruits arrive on a bus. When I got in the face of one of the female "recruits" she just sat down and began to cry. Shooting stopped so the director could remind the young lady I was only acting, and when she had regained her composure we did it again.

Later on I received a call from Fox asking if I could train two of their actors to fold a flag for an upcoming funeral scene. I said of course I could, but suggested they use two *real* Marines instead. A couple of the Marines in my detachment, Corporals Joe Hendrix and Eric Merkle, idolized Lee Ermey - not only could they repeat dialogue from *Full Metal Jacket* word for word, they could do it in *his* voice. It would be quite a treat for them to meet their idol, and as a bonus they would get to appear in a movie with him. The studio agreed it was a good idea, and promptly flew my two devil dogs up to the studio.

They were in hog heaven.

TRUE ACTION HEROES

MSG Detachment Tokyo, Japan

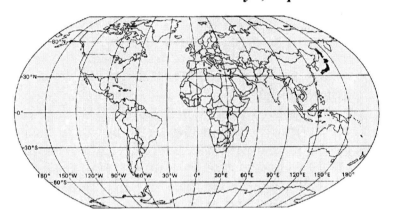

I said it before and I'll say it again - you never know <u>who</u> you might meet on the MSG program!

For some, he's the action hero that killed the alien in *Predator*. To others, he's the "Terminator." To California residents, he's Governor Schwarzenegger. But to Arnold Schwarzenegger, the men of the U.S. Marine Corps are the true action stars.

Schwarzenegger expressed that sentiment Friday night at the New Sanno Hotel in Tokyo, where he was the surprise special guest of the Marine Security Guard Detachment's Marine Corps Ball. The Governor, who has been in Japan on a trade mission, was invited to the celebration by Howard Baker, the U.S. ambassador to Japan, and the U.S. Embassy Marine Security Guard detachment.

"When I heard you had this celebration here today, I told

the Ambassador that there is *no way* I would miss it," Schwarzenegger told the crowd. "It's great to be here with the greatest of the great, and the strongest of the strong. I am honored to join you tonight, to celebrate the 229th anniversary of the birth of the Marine Corps."

California Governor (and action hero) Arnold Schwarzenegger.

During his speech to the more than one hundred Marine Corps Ball guests, the Governor spoke reverently about the Marine Corps' past. "You have always been there on the front lines fighting to protect America, and I know you have a great history of kicking some *serious* butt. The only reason why people become so successful in America, and the only reason why this country is so successful, is because of you brave men and women right here, you are the ones who protect the liberty," he said. "You are protecting this great country of America, and defending the American dream. I want to thank all of you for the great work you are doing.

You are doing a fantastic job of protecting our nation and safeguarding the stability of the Western Pacific." Schwarzenegger closed by saying, "I only *play* an action hero, but you people are the *true* action heroes!"

ZONA ROSA

MSG Det San Salvador, El Salvador

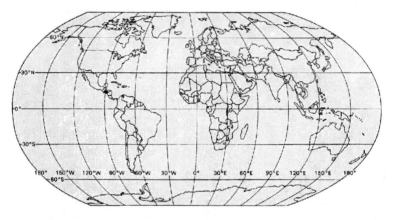

An MSG, or any American overseas for that matter, should never, <u>ever</u> let his guard down - period.

At around 8:30 PM on 19 June 1985, six of the Marines who were responsible for security at the United States Embassy sat down at an outside table at Chili's Restaurant in the area known as the "Zona Rosa" in the San Benito district of San Salvador, El Salvador. They were regular customers known to the owners of restaurants and cafes in the area, and to those who worked there. They used to go there in groups whenever they were off duty. The Marines were attached to the United States Embassy Security Guard Detachment and, although not in uniform, were easily identifiable as United States Marines by their haircuts, clothing, and security radios. After awhile, two of them left the group and went to sit down at a table in the Flashback Restaurant a few yards away from their companions at Chili's.

At around 9 PM a white pick-up truck with dark stripes parked outside the La Hola Restaurant. A group of some seven individuals got out and walked over to Chili's and, without warning, fired a volley of shots at United States Marines Thomas Handwork, Patrick R. Kwiatkoski, Bobbie J. Dickson and Gregory H. Weber. The Marines were in civilian clothing, and there is no evidence that they were carrying weapons.

The bodies of Marines Thomas Handwork, Patrick R. Kwiatkoski, Bobbie J. Dickson and Gregory H. Weber about to leave El Salvador enroute to the United States.

One witness said the Marines had been approached by a young man who briefly spoke with them and then bicycled away. Ten minutes later, at about 2100, ten men wearing camouflage shirts and caps and riding in a light-colored pickup truck arrived in front of a group of four adjacent sidewalk cafes. The patrons of the cafes dismissed the group as a military patrol conducting a search or a document check. The truck parked on the street in front of Chili's where the Marines were seated. The attack element jumped from the truck and turned toward the customers while others from their group deployed to adjacent security positions. The attackers moved directly toward the table where the Marines were seated while firing on full automatic with U.S. M16s, German G3s, and Israeli Uzi submachine guns. Fire was initially directed at the Marines, but was then turned against other patrons. One Marine was chased into "Las Pizzas" cafe and killed. Several other armed men provided security by directing gunfire towards the Brazilian Embassy across the street.

On June 21 a juvenile delivered an envelope to the Salvadoran National Police containing a communiqué he said he found in a telephone booth. The envelope contained a note from a Salvadoran guerrilla group called the 'Central American Revolutionary Workers' Party' (the Partido Revolucionario de los Trabajadores Centroamericanos or "PRTC"), which was one of the guerrilla groups constituting the 'Farabundo Marti National Liberation Front' (FMLN). The communiqué stated that the FMLN claimed responsibility for the "annihilation attack against American military advisors."

From that communiqué it is apparent that these organizations do not have the ability (or the interest) to differentiate between "advisors" and Marine Security Guards. They only know they killed some Americans. It further shows that danger lurks around every corner, and we can *never* let our guard down for a moment.

MOSCOW STATION

MSG Det Moscow, USSR

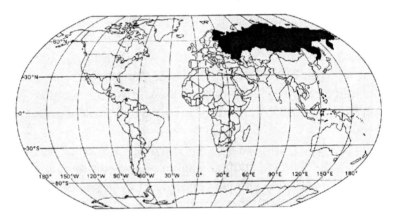

You have read about both the dangers and rewards of the MSG program up to this point, but no book about the subject would be complete (unfortunately) without mention of the "Moscow Station" incident. It was an aberration to be sure, but things like this can and do happen and must be guarded against. If you don't already know the story, Sergeant John Clayton Lonetree was a Marine Security Guard at the U.S. Embassy in Moscow from September 1984 until March 1986. Later in 1986, in Vienna, he was placed under detention after he acknowledged his involvement with a female KGB officer, Violette Seina, who had previously been a telephone operator and translator at the Embassy in Moscow. Soon after their relationship began, Seina introduced Lonetree to her "Uncle Sasha" who was later identified by U.S. intelligence as being a KGB agent. It was alleged at the time

that Lonetree had a sexual liaison with Seina, and had in fact allowed Soviet agents after-hours access to the Embassy. In December of 1986, Lonetree turned himself in to authorities at the U.S. Embassy in Vienna, Austria, where he was stationed. Also arrested and charged with collaboration was Corporal Arnold Bracy who was alleged to have been romantically involved with Soviet women. As the investigation proceeded, five other MSGs were detained on suspicion of espionage, lying to investigators, or improper fraternization with foreign nationals. Lonetree was tried on thirteen counts, including espionage. Among these counts were charges that he conspired with Soviet agents to gather names and photographs of American intelligence agents, to provide personality data on American intelligence agents, and to provide information concerning the floor plans of the U.S. Embassies in Moscow and Vienna. On 21 August 1987 he was convicted of espionage by a military court, and sentenced to thirty years imprisonment, a fine of $5,000, loss of all pay and allowances, reduction to the rank of private, and a dishonorable discharge. Espionage charges against Bracy and all of the other Marines have since been dropped. According to reports in late 1987, intensive investigations led to the conclusion that the former guards did not, as earlier believed, allow Soviet agents to penetrate the Embassy in Moscow. Be warned, however – the whole mess began when one of the MSGs was simply asked to provide an unclassified phone book (which was not supposed to be given out). Most of the following was derived from articles in Time magazine and other publications.

The "Marine spy scandal" that started with a lonely U.S. embassy guard confessing he had succumbed to the charms of a beautiful Soviet receptionist in Moscow eventually escalated into what appeared to be one of the most serious sex-for-secrets exchanges in U.S. history. Not only had the Marine's partner been charged with helping him let Soviet agents prowl the embassy's most sensitive areas, but later a third Marine sentinel was accused of similar offenses. A fourth Marine, stationed at the Brasilia embassy, was subsequently taken to Quantico, Virginia for grilling about espionage. Several others were recalled from Vienna. And then more accusations of spying were filed.

Sergeant John Weirick spread the contamination to the U.S. consulate in Leningrad, where Weirick too permitted KGB agents to enter at the urging of a Soviet woman. That prompted the State Department to cut off all electronic communications with the consulate and order the recall of the six-man Marine contingent in Leningrad, as it had earlier recalled the twenty-eight-man detail at the Moscow embassy. Ominously, Weirick's collaboration with the KGB occurred in 1982, four years *earlier* than the Moscow treachery, indicating a long-standing security breach.

Weirick, who was arrested at the Marine Corps Air Station in Tustin, California, later served at the U.S. embassy in Rome, where other members of the Marine Detachment were questioned. As more than seventy agents from the Naval Investigative Service set about the numbing task of locating, grilling and polygraphing every one of the more than two hundred Marines who had served at the Moscow and East European embassies during the previous the decade, they discovered that all but a few of the first fifty they quizzed flunked questions about fraternizing with local women.

The proud U.S. Marine Corps, whose often heroic Leathernecks had long boasted of being nothing short of the best, was confounded. "We've now got to operate on the thesis that this is possibly an endemic problem in the

Marines," said a senior officer at the Corps' Washington headquarters. Declared another officer: "I'm stupefied, flabbergasted. We just never thought something like this could happen."

So battered was the Corps that Major General (and future Commandant) Carl Mundy resorted to an otherworldly defense when grilled by a House committee when he paraphrased the Marines' Hymn: "If you look on heaven's scenes, you'll find the streets are guarded by United States Marines."

As members of Congress expressed bipartisan outrage, President Reagan ordered Secretary of State George Shultz to protest the Soviet penetration of the U.S. embassy directly to Foreign Minister Eduard Shevardnadze when the two begin talks on a treaty to eliminate intermediate-range missiles in Europe. The President also set in motion half a dozen seemingly redundant investigations into embassy security.

But Reagan and Shultz would not accede to a Senate resolution calling for the Secretary to postpone his Moscow trip until security problems were resolved. Shultz conceded that the espionage threw a "'heavy shadow" over U.S.-Soviet relations, but Reagan declared, "I just don't think it's good for us to be run out of town." The Administration's priority, he told the Los Angeles World Affairs Council, is the "pursuit of verifiable and stabilizing arms reduction." The President even repeated his invitation to Soviet Leader Mikhail Gorbachev to come to the U.S. for a summit: "The welcome mat is still out."

Nevertheless Shultz, who accepted ultimate chain-of-command responsibility for the embassy problems, was in the difficult position of flying into Moscow accompanied by a special communications van to help replace the compromised facilities at the U.S. embassy. Even the "Winnebago," as it became known, may not have protected him. When checking the supposedly secure trailer in Washington for emissions at frequencies believed used by the sophisticated Soviet bugs planted in the U.S. embassy, technicians found, according to one, that the Winnebago "radiated like a microwave." Similar vans have long accompanied U.S. Presidents abroad, raising the possibility that their communications back to Washington may have been overheard.

The pervasive spy scandal was an embarrassment for an Administration that had proclaimed its security consciousness and advocated wider use of lie-detector tests among federal employees to protect secrets at home. Administration officials, and the State Department in particular, displayed a curiously casual attitude toward the vulnerability of its embassies to Communist snooping. Washington was aware of the problem: White House sources said the issue had been raised repeatedly for years. Before the Geneva summit in November 1985, the senior White House staff received a National Security Council briefing on the Soviet Union's techniques for electronic surveillance and, for what was a prudish culture, its blatant use of sexual entrapment. The President's Foreign Intelligence Advisory Board issued at least three reports on the subject and personally briefed Reagan on the vulnerability of the Moscow embassy - but all these initiatives died, White House aides contended, amid bureaucratic sluggishness and even outright resistance on the part of the State Department.

Indeed, the high-tech proliferation of miniaturized, and in some cases virtually undetectable, eavesdropping devices seemed to have promoted a defeatist 'we'll-have-to-live-with-bugs' attitude. "Our security people have always looked upon our buildings as loaded with bugs," explained a former foreign service officer, who dismissed sexual entrapment as just another professional hazard. Such complacency may have contributed to what a high State Department official described as a "first-class mess."

It took months to assess the precise damage inflicted by the spying, but a senior White House official declared early on, "These cases taken together are likely as significant as the worst hits of the past." They were at least as serious, he claimed, as the Navy's Walker-family spy ring, the sale of secrets by the National Security Agency's Ronald Pelton, and the defection of former CIA Employee Edward Howard. The damage extended far beyond matters related to the Soviets. The Moscow embassy was on the distribution list for a wide range of foreign policy material, including details of U.S. negotiating positions in the Geneva arms talks, background on Nicaragua policy, Middle East affairs and relations between the U.S. and its allies. The CIA had its own communications facilities in Moscow, and the agency assumed that those too were compromised.

As the scandal spread, U.S. diplomats were rendered almost mute in their enclaves in Eastern Europe, reduced to writing sensitive messages in longhand. Even in non-Communist countries, the uncertainty of who might be listening turned U.S. envoys into near paranoids. On a trip in Southern Africa, Assistant Secretary of State Chester Crocker refused to send any reports to Washington until he could do so personally. "It's incredible the impact of this on all of us," said a State Department official. In an age of wondrous globe-spanning communications, the superpower that pioneered the technology found its creations turned against it.

The treasonous acts attributed to the Marine guards were

bad enough, but most of Washington was also belatedly aroused by the long-known and festering problem of the new U.S. embassy compound in Moscow which was nearing completion when work was halted in 1985. Built from prefabricated sections produced off the site - and out of sight of any U.S. inspectors - the chancery, not surprisingly, was found riddled with embedded snooping gear. Charged Texas Republican Congressman Dick Armey: "It's nothing but an eight-story microphone plugged into the Politburo."

Reagan vowed that the Soviets would not be permitted to occupy their new embassy on Mount Alto in Washington until security could be assured for the U.S. in its new Moscow quarters. He conceded that the red-brick U.S. chancery, whose walls were already water-stained because of its unfinished roof, may have been so bug-ridden that it would have to be demolished. The entire complex, which included 114 occupied residential units and recreational facilities, had been budgeted at $89 million. The cost when it was finished, apart from the electronic cleansing, ended up being $192 million.

On Capitol Hill, Republican Senators Robert Dole and William Roth introduced a tough package of anti-espionage measures that would require the President to negotiate a new site for the U.S. embassy in Moscow. If the Soviets did not provide such a site, including security guarantees, they would have been required to vacate their entire new Mount Alto compound in Washington.

As Republicans took the lead in berating the Administration for the security fiasco, Indiana's Senator Richard Lugar released a report compiled by the Senate Foreign Relations Committee while he was chairman. It charged the State Department with "poor management and coordination" in protecting embassies against Soviet penetration. Lugar then called on the White House to suspend the construction of new embassies in Bulgaria, Czechoslovakia, East Germany, Hungary and China until the embassy security investigations were completed.

Congressional anger was dramatized by a showboating but nonetheless revealing jaunt to Moscow by Democratic Congressman Dan Mica of Florida, chairman of the House Subcommittee on International Operations, and its ranking Republican, Maine's Olympia Snowe. Accompanied by a TV crew and four aides, they barged into the old embassy around midnight and approached the Marine guard on Post One. "May I see some ID, please?" the sentry asked politely. He examined passports, logged names, made a phone call, then issued visitors' ID cards. "Is this the place where Lonetree worked?" Snowe asked an embassy official. She of course was referring to Sergeant Clayton Lonetree, the first Marine to be arrested. The official hesitated, then offered a shrewd answer: "Er, in principle, yes." After a two-hour tour of the building and two days of interviewing, the legislators proclaimed the embassy not only "grossly inadequate for security purposes" but a "firetrap." Back in the U.S., Mica was blunter. "It's an absolute security disaster," he told *Time.* Ever since Lonetree was arrested, he said, embassy personnel had been communicating secret information in writing, often on children's erasable slates. Even then they shielded their messages from suspected hidden cameras. Any notes on paper were promptly shredded.

The embassy's security "bubble" and its massive vault had been declared off-limits to U.S. officials for classified conversations since those areas were broken into by Soviet agents. Two new secure rooms were hastily erected for Shultz's use, one of them described by Mica as similar to a "walk-in cooler, eight feet by ten feet, each with a folding table and a dozen chairs." Surprisingly, blueprints for those new rooms had been posted openly on an embassy wall. The cost of clearing bugs and replacing compromised gear was more than $25 million.

After talking to a third of the twenty-eight Marine guards, whose replacements had been held up by Soviet delays in issuing new visas, Snowe found them "depressed, humiliated, surprised, angry." Many, she said, realized that there had been a "total breakdown in discipline." Security was lax and "everybody at the embassy knew it," charged Snowe. Part of the blame had to fall on Arthur Hartman, the Ambassador who left the post in February of that year.

While admitting some of their own failures, the Marines claimed they were being used as scapegoats for the lackadaisical attitude toward security shown by diplomatic personnel. Snowe said the Marines had reported finding 137 violations that year, including open safes and classified papers left exposed. Conceded a Washington source: "One unfortunate result of this mess will be further alienation of the Marines and the State Department types."

Some MSGs insisted that the embassy civilians were also guilty of fraternizing with Soviets. The rules against fraternization in Soviet bloc nations required all embassy employees, from the Ambassador to the Marine guards, to report any "contact" with a national of the host country in an "uncontrolled" situation. The rule-breaking made it easy for Violetta Seina, a former receptionist at the U.S. Ambassador's residence, to seduce Lonetree into letting the KGB enter the embassy. He claimed to have met her on a Moscow subway, although she attended the annual Marine ball at the embassy. Galina (her last name was not revealed), the cheerful Soviet cook at Marine House, also had easy access to Corporal Arnold Bracy, the guard she befriended.

According to Navy investigators, Lonetree's pride in his love affair with Seina led indirectly to his arrest. In one account, he and an unidentified corporal later visited Stockholm together and went on a drinking binge in the Marine quarters at the U.S. embassy there. The booze loosened Lonetree enough for him not only to describe his passion for Seina, but also to reveal hints of a KGB connection. Later, when the two drinking buddies met in Vienna - where Lonetree was posted after Moscow - they enjoyed another blast. This time Lonetree mentioned Bracy's involvement as well.

Weirick also was alleged to have been led to the KGB by several women he encountered while stationed at the Leningrad consulate. He left Leningrad in 1982 and was transferred to Rome, where investigators contend he bragged to a colleague of having earned some $350,000 from the Soviets.

Spying is an old and nasty game among rival nations. The key issue in the sad Marine espionage scandal was not whether the Soviets had broken some unwritten rule of civilized snooping, or what American agents had done to them. A more relevant question was just why American Marines and State Department officials had permitted the Soviets to compromise U.S. security so thoroughly - and so easily.

IS THIS PART OF MY
Job Description?

MSG Detachment Cairo, Egypt

Neil R. Huff

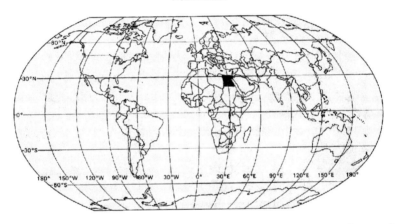

MSG duty can best be described at long stretches of boredom, punctuated by moments of stark terror (as was the case in Nairobi and Dar es Salaam) or incredible opportunity (as was the case here). You never know which one you will get – so train yourself to think out of the box.

In the fall of 1956 Britain and France attacked Egypt when talks broke down with General Nasser over Egypt's nationalization of the Suez Canal. Israel, scenting blood, quickly joined in the assault, attacking across the Sinai Desert toward Egyptian positions along the canal. Anglo-French naval, air and landing forces hit swiftly on October 31. Egypt, while aware of a possible Anglo-French attack,

278

was stunned by the Israeli invasion across the Sinai. Egypt was to be utterly overwhelmed in the brief but destructive war that ended with the canal blocked and occupied by the enemy, several Egyptian cities blasted, her military forces routed, and vital forts guarding the Canal lost to Israel.

Americans in Egypt were warned days in advance of a possible conflict and had been advised by the Embassy to leave the country. Naturally enough, very few heeded this good advice and when the air, land and sea attacks began, about 2300 Americans - tourists and residents - were still there and woefully unprepared for an emergency evacuation. For Americans in Cairo, the war began in the early hours of November 1. We were awakened at the Marine House by the drone of fighter bombers, the distant thudding of bombs and the nearer banging of AAA batteries on the outskirts of the city.

We off-duty Marines rushed to the rooftop of our house and watched AAA bursts dot the sky, but hopelessly behind and below the carrier-based jets. It was the start of a day to remember, for besides dealing with the problems of a war breaking out about our heads, we had to evacuate more than two thousand American citizens. Among the evacuees was the prettiest American girl in Cairo - a lady the writer was fated to marry forty-three years later. By 0900 two or three thousand people were milling around in front of the embassy, blocking traffic and sending the police into whistle blowing, arm waving delirium. They were mainly U.S. citizens, with their pets and children's nannies and piles of luggage in tow. Compounding the confusion were several hundred frightened nationals of a score of other countries literally fighting to get into the U.S. Embassy grounds and hopefully onto our evacuation buses.

It was a scene of utter bedlam. There were only eight

Marines and a scaled down embassy staff numbering about fifty to deal with the chaos of evacuation. We had to burn seventy-five years worth of classified documents, prepare the embassy for possible mob assaults, and manage all the complications of dealing with a host government fighting for its existence.

Then there were the animals. Every American family owned at least one pet, and many people brought them to the embassy that morning despite having been told to leave them at home. They ranged from lizards to birds to donkeys. All of them - dogs, cats, birds and the donkey, were left behind and had to be roped and staked out - by the Marines, naturally - on the embassy lawn.

There must have been at least a hundred and fifty of all sorts of livestock left in the care of the embassy. As a thoughtful addition to the general commotion that day the Egyptian government, in a moment of pique, withdrew our police protection.

The black uniformed cops climbed into their truck just as a mob was forming half a block away. I remember one waggish cop pointing toward the mob, grinning, and sawing a forefinger back and forth across his throat. But that is another story. Before we off-duty Marines finished dressing that morning we heard the clatter and rumble of tracked vehicles in the streets. To our astonishment, these turned out to be Egypt's new Soviet-supplied tanks. When western aid was halted following Nasser's nationalization of the Suez Canal, Egypt had turned to the Soviet Union. They, of course, leapt at the chance of gaining a foothold in the Middle East. When Nasser asked for military aid the Soviets obliged, quickly rushing in all types of military hardware - including their latest main battle tanks. The monster machines chewing up the pavement in Garden City included

Joseph Stalin II's, the newest tank in the Soviet arsenal. General Nasser wisely refused to commit his newly equipped armored divisions to the unequal battles then being waged, and chose instead to hide these forces in Cairo under the trees along the shady streets of Garden City and Zemalik.

As we watched the clanking parade passing our house, one enormous JS II broke ranks, pulled over and halted under the trees next to our wall. The meter-wide treads flattened the sidewalk and nudged a dent in the brick base of our wall of iron pilings. As we watched, fascinated by this huge machine - half again as big as any tank in the West - the hatches flew open and out popped the crew, one after the other, smallish men who kept coming out like dwarfs from a circus car.

Soviet-made Joseph Stalin II tank.

Finally there were six or seven of them lined up in front of a nervous, yelling sergeant. They gave us surly looks as we gathered near the gate and watched the show while waiting for the embassy cars to pick us up. At that point the average Egyptian had no idea who was attacking their country. For all these poor fellows knew, the planes overhead might have been ours! Little wonder they glared at us. Later that day, when we were ordered back to the Marine House for meals, we found that the tank crew had gotten out a couple of tents in the meantime and erected them on the sidewalk. They appeared to be settling in for a long stay. It was clear at the outset that the sergeant had instructed his men to refrain from consorting with the foreigners.

They may have given us dirty looks, and mumbled insults, but our house boys soon enough meandered out to talk with the tankers and inform them that we were the U.S. Marines who guarded the American Embassy. Very quickly the tankers rigged up a few blankets and tried their best to hide from view what they thought were the most important components of their tank. They knew better than we did at that point that their Soviet tank was new and innovative. One thing that *did* impress us was the enormous gun jutting from the inverted frying pan turret of that big tank. Nothing could conceal that 120 mm muzzle. It looked like a telephone pole!

Our NCOIC lost no time telling the Army attaché at the embassy that this JS II was parked almost in our yard. I'm unsure what he was told in detail, but soon enough the off-duty Marines were working hard to break through the reserve of the Egyptian tankers and befriend them. They began by inviting them inside our yard to put up their tents on the lawn.

We allowed them to draw water from our taps, soon they were accepting cigarettes, and within another twenty-four hours all of them including the hard-bitten sergeant were drinking beer with us and eating from our kitchen - with all of us loudly condemning the Brits, French and Israelis for their cowardly attack. In fact, we all were quite sympathetic toward the Egyptians, who were and remain extremely kind and generous people, and our condemnations were genuine enough.

This friendly fraternization continued, and soon the crew was so disarmed that they no longer bothered to keep one of their number posted on guard by the tank while the rest of the crew lounged and drank in the Marine House bar. It was on the second or third night that a fellow from one of the attaché offices came over with a Minox camera and flash. It was the work of a minute or two for the Army Attaché's man, assisted by one of the Marines, to climb into the tank and photograph the interior.

As far as I can recall, the Marines never learned the outcome of the photo session. I assume the attachés got the pictures they wanted. While it wasn't any part of our job to spy on the host country, the opportunity to photograph the JS II was simply too good to pass up. In the ordinary course of an Embassy Guard's duties any activity remotely suggestive of intelligence gathering always has been strictly forbidden, for nothing would endanger the MSG presence faster than suspicion by a host government that the Marines were engaged in espionage. For myself, I never felt we had spied against the Egyptians.

This happened during the Cold War and the Soviets, whose tanks these were, were our opponent. That thought always eases my conscience whenever I recall Egypt, the Suez war, those hungry soldiers, and their bloody great tank squatting on our front steps.

NICE KITTY

MSG Detachment Brazzaville, Congo

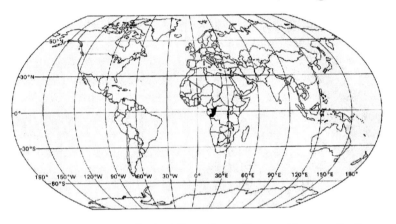

This section is called "You Just Never Know..." for a reason, and this story is a great example of what can happen when you least expect it. It also shows why the State Department generally gets out of our way in a crisis!

One of the strangest events to occur during my tour in the Republic of the Congo took place during the summer of 1993 while the post was under an evacuation order for all dependants and non-essential personnel. The city of Brazzaville had been a maze of armed barricades for several weeks, and a 1900 curfew was in effect in an attempt to limit the proliferation of running gun battles in the streets. One deceptively idyllic Sunday afternoon, while the guns were silent, some of the Embassy staff decided to take advantage of the lull and get out for a while.

One of those who ventured out was a communicator

named Ann who went to the local zoo along with a visiting Diplomatic Courier to feed the lions. The Brazzaville Zoo in no way resembles the San Diego Zoo since the animals are given scarcely enough food to survive, with the 'surplus' going to the families of the zookeepers. The courier, who was based in Frankfurt Germany, foolishly stepped over the low barrier that separates visitors from the cages and held out a piece of fish. When Ann called out to her she turned her head, and as she did so the lion decided that he would rather have *her* for lunch than the fish. He hooked the courier with his claw and pulled her arm through the bars of the cage - and proceeded to bite off most of her right shoulder as well as a chunk of her left forearm.

A well-fed cousin of the lion in Brazzaville.

A local employee of the chimpanzee project which is collocated with the zoo beat the lion with a stick in an attempt to get him to release his grip, and Ann got on her handheld radio and began to scream hysterically. I was at my residence - known as "Villa Gunny" - several blocks away when the initial call came in, and spent several minutes along with our Political Officer and the Marine on duty trying to get a coherent response out of her. Fortunately for the courier the employee who had wrested her from the lion's grasp had enough presence of mind to apply a tourniquet, and when we arrived we were able to take her by taxi to what passes for a hospital in Africa.

I then mobilized my detachment, and along with the Political Officer began rounding up surgical supplies and blood donors (the local supply was tainted with AIDS) via radio. The Embassy staff arranged for a medically equipped aircraft to come in from Johannesburg to fly the victim to South Africa, and later that night a military escort provided by the Congolese helped ensure our convoy got to the airport without incident despite the curfew.

Several articles were written in various publications about the incident, none of which were accurate, and the whole thing was largely forgotten save for the occasional lion joke. That is until the State Department, in a classic example of why it is one of the largest and least effective bureaucracies in the world, presented Ann with an award for heroism. We're still trying to figure that one out. Everyone involved from the blood donors to the Marine on Post One to the pilot of the medevac aircraft had done more than she, but if *anyone* truly deserved the label of hero it was the Congolese gentlemen who took immediate action and saved the courier's life.

MEGA

Marine Embassy Guard Association

The Marine Embassy Guard Association is exceptionally proud of the heritage of Marine Security Guards and their service to Country and Corps. Its purpose is to protect and preserve that heritage, and during difficult and hostile times around the world the members of MEGA keep America's Marine Security Guards close in their thoughts and hearts.

The Association is comprised of Marines who have served or are currently serving as Marine Security Guards since the inception of the program in 1948, and welcomes the participation of all Marine Security Guards everywhere, both active and inactive, and regardless of how large or small their contributions.

Find them on the worldwide web at:

www.embassymarine.org

ATTACKS ON THE U.S. DIPLOMATIC CORPS

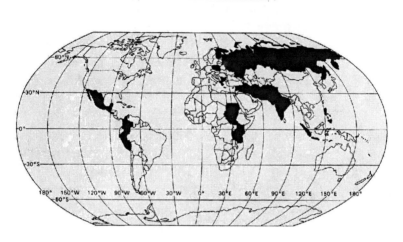

These are just some of the attacks on DOS targets protected by MSGs around the world:

September 2006: Attackers try to blow up the U.S. Embassy in Damascus, Syria but are thwarted by Syrian guards in a fierce gun battle. Three of the four attackers and a Syrian guard were killed.

March 2006: A suicide bomber kills a U.S. diplomat and three other people near the consulate in Karachi. The attack comes two days before President Bush is due to visit Pakistan.

December 2004: Four Saudi national guardsmen die when gunmen attack the U.S. consulate in Jeddah, Saudi Arabia. Three assailants also die.

July 2004: Four local civilians and three suicide bombers die in Tashkent, capital of the central Asian state of Uzbekistan, in simultaneous blasts outside the U.S. and Israeli embassies.

May 2004: A policeman is killed and thirty-two other Pakistanis injured in a bomb attack on the residence of the U.S. consul in Karachi.

February 2003: A gunman kills three Pakistani policeman guarding the U.S. consulate in Karachi. The assailant is also killed.

June 2002: A car bomb exploded near the U.S. Consulate and the Marriott Hotel in Karachi, Pakistan. Eleven persons were killed and fifty-one were wounded, including one U.S. and one Japanese citizen. Al Qaida and al-Qanin were suspected.

March 2002: A car bomb exploded at a shopping center near the U.S. Embassy in Lima, Peru. Nine persons were killed and thirty-two wounded. The dead included two police officers and a teenager. Peruvian authorities suspected either the Shining Path rebels or the Tupac Amaru Revolutionary Movement. The attack occurred three days before President George W. Bush visited Peru.

January 2002: Armed militants on motorcycles fired on the U.S. Consulate in Calcutta, India, killing five Indian security personnel and wounding thirteen others. The Harakat ul-Jihad-I-Islami and the Asif Raza Commandoes claimed responsibility. Indian police later killed two suspects, one of whom confessed as he died.

December 2000: A bomb exploded in a plaza across the street from the U.S. Embassy in Manila, injuring nine persons. The Moro Islamic Liberation Front was likely responsible.

August 1998: A bomb exploded at the rear entrance of the U.S. Embassy in Nairobi, Kenya, killing twelve U.S. citizens, thirty-two Foreign Service Nationals (FSNs), and 247 Kenyan citizens. Approximately five thousand Kenyans, six U.S. citizens, and thirteen FSNs were injured. The U.S. Embassy building sustained extensive structural damage. Almost simultaneously, a bomb detonated outside the U.S. Embassy in Dar es Salaam, Tanzania, killing seven FSNs and three Tanzanian citizens, and injuring one U.S. citizen and seventy-six Tanzanians. The explosion caused major structural damage to the U.S. Embassy facility. The U.S. Government held Usama Bin Laden responsible.

February 1996: Unidentified assailants fired a rocket at the U.S. Embassy compound in Athens, causing minor damage to three diplomatic vehicles and some surrounding buildings. Circumstances of the attack suggested it was an operation carried out by the 17 November group.

September 1995: A rocket-propelled grenade was fired through the window of the U.S. Embassy in Moscow, ostensibly in retaliation for U.S. strikes on Serb positions in Bosnia.

February 1991: Three Red Army Faction members fired automatic rifles from across the Rhine River at the U.S. Embassy Chancery in Bonn, Germany. No one was hurt.

January 1991: Iraqi agents planted bombs at the U.S. Ambassador to Indonesia's home residence and at the USIS library in Manila.

January 1990: The Tupac Amaru Revolutionary Movement bombed the U.S. Embassy in Lima, Peru.

June 1988: The Defense Attaché of the U.S. Embassy in Greece was killed when a car-bomb was detonated outside his home in Athens.

June 1985: Members of the FMLN (Farabundo Marti National Liberation Front) fired on a restaurant in the Zona Rosa district of San Salvador, killing four Marine Security Guards assigned to the U.S. Embassy and nine Salvadorean civilians.

March 1984: The Islamic Jihad kidnapped and later murdered Political Officer William Buckley in Beirut, Lebanon. Other U.S. citizens not connected to the U.S. government were seized over a succeeding two-year period.

April 1983: Sixty-three people, including the CIA's Middle East director, were killed and 120 were injured in a 400-pound suicide truck-bomb attack on the U.S. Embassy in Beirut, Lebanon. The Islamic Jihad claimed responsibility.

November 1979: After President Carter agreed to admit the Shah of Iran into the U.S., Iranian radicals seized the U.S. Embassy in Tehran and took sixty-six American diplomats hostage. Thirteen hostages were soon released, but the remaining fifty-three were held until their release on January 20, 1981.

February 1979: Four Afghans kidnapped U.S. Ambassador Adolph Dubs in Kabul and demanded the release of various "religious figures." Dubs was killed, along with four alleged terrorists, when Afghan police stormed the hotel room where he was being held.

August 1974: U.S. Ambassador to Cyprus Rodger P. Davies and his Greek Cypriot secretary were shot and killed by snipers during a demonstration outside the U.S. Embassy in Nicosia.

May 1973: U.S. Consul General in Guadalajara, Mexico Terrence Leonhardy was kidnapped by members of the People's Revolutionary Armed Forces.

March 1973: U.S. Ambassador to Sudan Cleo A. Noel and other diplomats were assassinated at the Saudi Arabian Embassy in Khartoum by members of the Black September organization.

ABOUT THE AUTHOR

Andy Bufalo retired from the Marine Corps as a Master Sergeant in January of 2000 after more than twenty-five years service. A communicator by trade, he spent most of his career in Reconnaissance and Force Reconnaissance units but also spent time with Amtracs, Combat Engineers, a reserve infantry battalion, and commanded MSG Detachments in the Congo and Australia.

He shares the view of Major Gene Duncan, who once wrote, "I'd rather be a Marine private than a civilian executive." Since he is neither, he has taken to writing about the Corps he loves.

CPSIA information can be obtained at www.ICGtesting.com
Printed in the USA
269472BV00005B/3/P